OVERCOMING SELF-SABOTAGE

OVERCOMING SELF-SABOTAGE

Tamika D. Roberson

TABLE OF
CONTENTS

INTRODUCTION

In a world that often equates success with self-worth, the fear of failure can become a formidable adversary. The pressure to excel, to meet expectations, and to avoid mistakes at all costs can lead to a paralyzing anxiety that stifles potential and creativity. "Overcoming Self-Sabotage" addresses this pervasive issue, offering a trans-formative perspective on failure and personal growth.

"Overcoming Self-Sabotage" is a guide for those who find themselves trapped in a cycle of self-doubt and apprehension. It offers insights into understanding the roots of these fears, shedding light on how societal norms and personal experiences shape our perceptions of success and failure. The book is woven with relatable anecdotes and practical advice that demystify the concept of failure, illustrating how it can be a powerful catalyst for growth and self-discovery.

You will explore the psychological underpinnings of anxiety and self-doubt, gaining a deeper understanding of how these emotions can derail ambitions and hinder personal development. "Overcoming Self Sabotage" emphasizes resilience as a crucial skill in navigating life's uncertainties. By building mental fortitude and self-assurance, you can learn to approach challenges with courage and optimism. The book encourages you to redefine success on your own terms, fostering a mindset that values progress over perfection.

With compassion and clarity, "Overcoming Self Sabotage" invites you to confront your fears head-on, to dismantle the barriers that hold you back, and to emerge more confident and capable. This book is not just about conquering your fear of failure; it is about embracing the full spectrum of human experience, recognizing that setbacks are not reflections of inadequacy but rather steppingstones to a more resilient self.

One

UNDERSTANDING THE FEAR OF FAILURE

Manifestation of Fear

F ear is not just an emotion; it is a powerful force that can shape our lives in ways we often fail to recognize. It lurks in the shadows of our minds, whispering doubts and insecurities, convincing us that failure is the end rather than a steppingstone. But what if we could shift our perspective and see fear not as a barrier but as a catalyst for growth?

Consider the moments when fear has gripped you—those instances where your heart raced, and your palms turned clammy. These are not signs of weakness but indicators of your humanity, reminders that you are alive and engaged with the world around you. Fear, in its essence, is a natural response designed to protect us from harm. Yet, when it manifests in aspects of our lives where the stakes are not life-threatening, it can become a formidable adversary, one that stands between us and the realization of our true potential.

The manifestation of fear often begins subtly, with a quiet voice in our heads that questions our abilities and worth. It tells us that taking risks is foolish, that the possibility of failure outweighs the potential for success. This voice, though persistent, is not infallible. It thrives on our hesitation, feeding on the anxious energy we unwittingly provide. However, understanding this dynamic is the first step in reclaiming control.

Imagine fear as an artist, painting vivid pictures of worst-case scenarios that play out in our minds. It crafts scenes of public humiliation, of dreams dashed upon the rocks of reality. Yet, these are nothing more than illusions, projections of what might be, rather than what will be. By recognizing this, we can begin to dismantle the narratives that fear constructs, replacing them with stories of resilience and courage.

What if, instead of succumbing to fear's narrative, we reframed our understanding of failure? What if we saw it as an opportunity to learn, to grow, and to refine our approach? Failure is not a mark of shame but a badge of experience, one that speaks to our willingness to push boundaries and challenge the status quo. By embracing this mindset, we open ourselves to a world of possibilities, where fear is no longer an insurmountable obstacle but a companion on our journey.

In the grand tapestry of life, fear is but one thread. It can weave a pattern of limitation, or, when harnessed effectively, a design of empowerment. The choice is ours to make. We can allow fear to paralyze us, to keep us confined within the walls of our comfort zones, or we can choose to confront it head-on, to use its energy to propel us forward.

The manifestation of fear is a call to action, a challenge to redefine our relationship with failure. It invites us to step beyond the shadows and into the light of possibility, to rewrite the scripts that have been handed to us and author new ones that reflect our capacity for greatness. By acknowledging fear and its manifestations, we take the first step toward a life unburdened by shame, where failure is not feared but embraced as an integral part of our journey toward success. Let us not shrink in the face of fear but rise, empowered and unashamed.

Silencing Self-Doubt

Self-doubt is an insidious companion, whispering that you're not good enough, that your failures define you, and that success is reserved for others. It is a shadow that clings to your aspirations, trying to eclipse your potential. Yet, the truth is that self-doubt is a natural human emotion, one that can be transformed from a paralyzing force to a catalyst for growth.

Imagine a world where failure is not shunned but celebrated as a steppingstone to success. In this world, self-doubt loses its power. It becomes a mere signal, alerting you to areas for growth and learning rather than a verdict on your worth. To reach this place, you must first recognize that self-doubt is not the enemy. Instead, it is a protective mechanism that, when understood and harnessed, can propel you forward.

To silence the voice of self-doubt, you must first confront it. Acknowledge its presence without judgment. It's okay to feel uncertain or inadequate at times; these feelings do not define you. They are transient, like clouds passing over the sun. Acknowledge them, and then let them drift away, revealing the brilliance of your true capabilities.

Next, challenge the narratives that fuel your self-doubt. Often, these stories are rooted in past experiences or societal expectations. They are not immutable truths, but interpretations shaped by fear and insecurity. Reframe these narratives by focusing on your accomplishments, however small they may seem. Each success, each moment of resilience, is a testament to your strength and potential.

Surround yourself with voices that uplift and empower. Seek out mentors, friends, and communities that encourage growth and celebrate failure as a natural part of the learning process. Their support and belief in your abilities can be a powerful antidote to self-doubt. They remind you that you are not alone in your struggles and that others have walked this path before you, emerging stronger and more confident.

Set realistic goals and break them down into manageable steps. Self-doubt often thrives when the path ahead seems overwhelming. By creating a clear, achievable plan, you can focus on progress rather than perfection. Celebrate each milestone reached, no matter how small. These victories build momentum and confidence, gradually silencing the voice of doubt.

Practice self-compassion. Be kind to yourself, especially in moments of failure. Understand that everyone stumbles, and it is through these stumbles that we learn to walk with greater purpose. Treat yourself with the same kindness and understanding you would offer a dear friend. This compassion will fortify your resilience, making self-doubt a less formidable adversary.

Ultimately, silencing self-doubt is not about eradicating it entirely. It is about recognizing its presence, understanding its origins, and transforming it into a tool for personal growth. In doing so, you liberate yourself from the shackles of shame and step boldly into a future where failure is not feared but embraced as a necessary part of the journey towards success. By changing your relationship with self-doubt, you create a narrative of empowerment and resilience, one where you are not defined by your failures but by your courage to rise above them.

Overcoming the Anxiety

Anxiety is a formidable opponent, a silent yet persistent force that can overshadow our potential and shackle us to a cycle of self-doubt. To fail without shame, one must first confront and dismantle the anxiety that often accompanies the fear of failure. The key lies not in eradicating anxiety altogether but in transforming it into a catalyst for personal growth and resilience.

At its core, anxiety stems from the unknown, the uncertainty of outcomes, and the fear of judgment. It's a natural human response, yet it often spirals into a barrier that prevents us from taking necessary risks. Recognizing anxiety as a part of the human experience is the first step towards mitigating its power. It's essential to understand that anxiety is not a reflection of our capabilities or worth, but rather a signal that we are stepping into unfamiliar territory.

Shifting our perspective on anxiety involves an active choice to view it as an opportunity rather than a limitation. By acknowledging its presence, we can begin to dissect its components and understand the specific fears that fuel it. Is it the fear of rejection, the dread of criticism, or the anxiety of not measuring up? Identifying these underlying fears allows us to address them directly, transforming vague apprehension into manageable challenges.

Developing a mindset that embraces failure as a learning opportunity rather than a definitive end is crucial. This involves redefining success not as the absence of failure but as the ability to learn and grow from each setback. By consciously choosing to see failure as a steppingstone, we diminish its power to induce anxiety. Each attempt becomes a valuable lesson rather than a verdict on our abilities.

Practical strategies are essential in managing anxiety effectively. Mindfulness techniques, such as focused breathing and meditation, can help ground us in the present moment, reducing the overwhelming nature of anxious thoughts. Visualizing positive outcomes and rehearsing potential scenarios can also prepare us to face challenges with confidence and composure.

Equally important is cultivating a supportive environment. Surrounding ourselves with individuals who encourage risk-taking and view failure as a natural part of the growth process can significantly reduce anxiety. Sharing experiences and discussing fears openly can demystify the concept of failure, making it a shared human experience rather than a solitary burden.

Moreover, setting realistic expectations and breaking tasks into manageable steps can alleviate the pressure that contributes to anxiety. By focusing on achievable goals, we can build momentum and confidence, gradually reducing the fear of failure. Celebrating small victories along the way reinforces our ability to overcome obstacles and fosters a sense of accomplishment.

Ultimately, overcoming anxiety involves a continuous process of self-awareness, reflection, and adaptation. It requires us to challenge our perceptions, confront our fears, and embrace uncertainty as a natural part of life. In doing so, we empower ourselves to fail without shame, transforming anxiety into a driving force that propels us towards personal and professional fulfillment. By making peace with our anxieties, we open the door to a world of possibilities, unencumbered by the chains of fear and self-doubt.

Building a Resilient Mindset

Imagine a world where failure is not a dead-end but a steppingstone, a world where setbacks are seen as the architects of success. This is the world we ought to strive for—a world where resilience is our greatest ally. To thrive in this realm, cultivating a resilient mindset is not just beneficial; it is essential.

The path to resilience begins with a radical shift in perception. Rather than viewing failures as personal shortcomings, consider them as valuable experiences that shape our character and fuel our growth. This shift demands a conscious effort to replace self-doubt with self-compassion. By treating ourselves with the kindness we would extend to a friend, we foster an environment where learning from mistakes becomes second nature.

In the face of adversity, the language we use can be transformative. Words are powerful tools that can either uplift or undermine our resolve. By choosing language that emphasizes growth and potential, we reinforce our ability to overcome challenges. Instead of saying, "I can't do this," try, "I haven't mastered this yet, but I'm learning." This subtle change in dialogue can be the catalyst for a resilient mindset.

Moreover, embracing flexibility in our thought processes allows us to adapt to unforeseen circumstances. Life is unpredictable, and rigid thinking can leave us vulnerable to the winds of change. By cultivating mental agility, we equip ourselves with the tools to pivot and persevere, turning obstacles into opportunities.

Another cornerstone of resilience is the ability to maintain perspective. When faced with failure, it is crucial to step back and assess the broader picture. Often, what seems like a monumental setback in the moment is but a minor detour in the grand scheme of life. By maintaining a sense of proportion, we can prevent temporary failures from overshadowing our long-term goals.

Resilience is also nurtured through the power of community. Surrounding ourselves with supportive individuals who understand the value of perseverance can provide a buffer against the sting of failure. These connections remind us that we are not alone in our struggles and that shared experiences can fortify our resolve.

In this journey, self-reflection becomes a vital practice. Regularly evaluating our responses to failure allows us to identify patterns and make conscious adjustments. This introspection, however, must be approached with a growth mindset, focusing on what can be learned rather than dwelling on perceived inadequacies.

Finally, resilience is fueled by a sense of purpose. When our actions are aligned with our values and aspirations, the motivation to persist in the face of failure becomes intrinsic. Purpose provides the drive to push through the difficult moments, transforming setbacks into steppingstones toward our ultimate vision.

In a world where failure is inevitable, resilience becomes our shield and sword. By cultivating a mindset that embraces growth, adaptability, and community, we not only navigate the challenges of life but emerge stronger and more determined. This is the essence of failing without shame—turning each stumble into a powerful leap toward success.

Psychological Impacts

When we discuss failure, it often conjures images of defeat and inadequacy, a perception deeply ingrained in our societal psyche. This perception doesn't just manifest in our thoughts; it deeply impacts our mental health, shaping the way we view ourselves and the world around us. It's crucial to acknowledge and understand these psychological impacts if we are to redefine failure and liberate ourselves from its oppressive grip.

The fear of failure is a formidable adversary, lurking in the shadows of our minds. It whispers insidious doubts, convincing us that a misstep is a reflection of our worth. This fear doesn't just inhibit risk-taking; it fosters anxiety and stress, creating a self-imposed prison where potential remains unrealized. This fear can paralyze decision-making, stalling progress and innovation, and leading to a cycle of avoidance and stagnation.

The guilt associated with failure is another psychological burden many carry. It becomes a relentless companion, reminding us of our perceived shortcomings and amplifying feelings of shame. This guilt is not just a fleeting emotion; it can morph into a pervasive state of mind, affecting self-esteem and mental resilience. It is a thief of joy, robbing us of the ability to appreciate our successes and learn from our mistakes.

Moreover, societal expectations compound these feelings, perpetuating a narrative where success is the only acceptable outcome. This expectation fosters a culture of perfectionism, where the pressure to excel becomes overwhelming. Perfectionism, in turn, breeds a fear of vulnerability, as admitting failure is seen as a weakness rather than a steppingstone to growth. This cultural stigma surrounding failure inhibits open dialogue, leaving individuals isolated in their struggles.

However, by re-framing our understanding of failure, we can begin to dismantle these psychological barriers. It starts with recognizing that failure is not an endpoint but a necessary part of the learning process. This shift in perspective allows us to view setbacks as opportunities for growth rather than definitive judgments of our abilities.

Embracing this mindset encourages a culture of resilience, where individuals are empowered to take risks and learn from their experiences. It fosters an environment where vulnerability is not feared but celebrated as a pathway to deeper understanding and connection. By fostering open conversations about failure, we can dismantle the stigma and cultivate a supportive community that values growth over perfection.

Moreover, psychological resilience is strengthened by developing a growth mindset, which emphasizes effort and learning over innate ability. This mindset not only reduces the fear of failure but also enhances motivation and perseverance. It encourages individuals to view challenges as opportunities for development, rather than threats to their self-worth.

In shifting the narrative, we can transform the psychological impacts of failure from sources of shame and anxiety into catalysts for personal growth and innovation. By embracing this transformation, we not only liberate ourselves from the shackles of societal expectations but also pave the way for a more compassionate and understanding society. In this light, failure becomes not something to be feared, but a powerful ally in our journey towards self-discovery and fulfillment.

Two

EMBRACING FAILURE AS A STEPPINGSTONE

The Stigma Surrounding Failure

F ailure is often perceived as a shadow cast over one's achievements, a blemish that taints the bright canvas of success. Society has painted failure in hues of negativity, associating it with inadequacy, incompetence, and even disgrace. This perception is not only pervasive but also damaging, stifling innovation and inhibiting personal growth. It's essential to challenge this narrative and redefine our understanding of failure.

Culturally, success is celebrated with pomp and circumstance, while failure is swept under the rug, whispered about in hushed tones. From an early age, individuals are taught to equate failure with personal shortcomings rather than viewing it as a steppingstone to learning and improvement. This framing of failure as something to be ashamed of creates an environment where people are afraid to take risks, fearing the judgment and ridicule that might follow if they do not succeed.

The fear of failure can be paralyzing. It can prevent individuals from pursuing their dreams, trying new things, or stepping outside their comfort zones. When failure is stigmatized, it becomes a barrier to creativity and innovation. Many of the world's most successful individuals have experienced significant failures before achieving greatness. Their stories often highlight how these setbacks were instrumental in their eventual success. Yet, these narratives are overshadowed by the prevailing belief that failure is something to be avoided at all costs.

This stigma surrounding failure is deeply embedded in educational systems, professional environments, and social structures. In schools, students are often penalized for mistakes rather than encouraged to learn from them. In the workplace, employees might hide their errors, fearing repercussions instead of discussing them openly to foster a culture of learning and improvement. Socially, people are quick to judge others based on their failures rather than acknowledging the effort and courage it takes to try and fail.

To dismantle this stigma, there needs to be a paradigm shift in how failure is perceived. It should be seen as an integral part of the learning process, a necessary component of growth. Failure can teach resilience, perseverance, and adaptability—qualities that are invaluable in both personal and professional realms. By reframing failure as a positive force, we can create a culture that encourages risk-taking and values the lessons learned from setbacks.

One approach to changing this mindset is through open dialogue about failure. Sharing stories of failures and the insights gained from them can normalize the experience and reduce the associated shame. Encouraging environments where mistakes are viewed as opportunities for learning can foster innovation and creativity.

Moreover, it's crucial to celebrate effort and perseverance alongside success. Recognizing the courage it takes to attempt something new, regardless of the outcome, can help shift the focus from the result to the process. By doing so, we can create a more supportive and understanding society that values growth and learning over mere achievement.

Ultimately, the challenge lies in redefining failure not as a reflection of one's worth but as a powerful tool for development. Embracing this perspective can liberate individuals from the fear of failure, empowering them to pursue their aspirations with confidence and resilience.

Redefining Success and Failure

As we navigate through life, the concepts of success and failure are often presented to us in stark black and white terms. Society has conditioned us to chase after predefined benchmarks, equating success with material wealth, prestige, and accolades, while branding failure as a state of inadequacy and shame. This rigid dichotomy creates undue pressure and stifles our potential to explore new paths and learn from our experiences.

Imagine a world where success is not measured by the trophies on your shelf or the zeroes in your bank account, but by the depth of your experiences, the lessons learned, and the resilience built along the way. The traditional view of success is limiting, a narrow corridor that negates the richness of human experience. To redefine success, we must first dismantle the notion that it is a destination to be reached. Instead, we should view success as a journey of growth, where each step, irrespective of its outcome, contributes to our personal and professional development.

Similarly, failure should not be shunned or feared. Instead of viewing failure as a blemish, imagine it as a badge of honor, a testament to your courage to venture beyond the familiar and attempt the unknown. Every failure is a step towards mastery, a vital component of any successful endeavor. When we redefine failure as a learning opportunity, we transform it from an end into a beginning, a chance to reinvent ourselves and refine our strategies.

The fear of failure often holds us back, paralyzing us in the face of potential risk. However, when we shift our perspective, recognizing that failure is not the antithesis of success but a part of it, we liberate ourselves from the shackles of self-doubt. This liberation is the key to unlocking creativity and innovation, fostering an environment where we can experiment and explore without the fear of judgment.

Redefining these concepts also requires a shift in our societal narratives. It calls for a collective effort to celebrate not just the end results but the process and the perseverance behind them. We need to applaud the courage to try, the resilience to rise after a fall, and the wisdom gained from each setback. This shift will encourage a culture where individuals are motivated to pursue their passions and take calculated risks, knowing that failure is not a dead-end but a detour to greater understanding and achievement.

In this new paradigm, success becomes a personal construct, unique to each individual. It is not dictated by external validation but defined by personal satisfaction, fulfillment, and growth. This perspective empowers us to pursue paths that resonate with our values and passions, leading to a more authentic and meaningful life.

By redefining success and failure, we not only free ourselves from societal constraints but also open the door to a world of possibilities. We cultivate resilience, foster innovation, and embrace the full spectrum of human experience. It's time to break free from the conventional molds and redefine what it truly means to succeed and fail, without shame or fear.

Learning from Setbacks

Failure often feels like a looming cloud, casting shadows over our achievements and aspirations. Yet, hidden within its depths lies a treasure trove of lessons waiting to be unearthed. The path to success is seldom a straight line; it is a winding road filled with unexpected turns, and sometimes, daunting setbacks. But what distinguishes those who ultimately succeed is their ability to extract knowledge from these experiences and use it as a driving force for growth.

When faced with setbacks, it's natural to feel a sense of defeat. However, it's crucial to redefine our perception of failure. Instead of viewing it as an endpoint, consider it a valuable feedback mechanism. Each setback provides insights into what went wrong and what could be improved. By analyzing these moments with an open mind, we can identify patterns, understand our weaknesses, and recalibrate our strategies.

Moreover, setbacks challenge us to develop resilience. Resilience is not merely the ability to bounce back but the capacity to adapt and thrive in the face of adversity. It is about cultivating a mindset that sees challenges as opportunities for personal development rather than insurmountable obstacles. By fostering resilience, we build the mental fortitude needed to navigate future challenges with confidence and determination.

Embracing setbacks also encourages innovation and creativity. When conventional methods fail, we are compelled to think outside the box and explore alternative solutions. This process often leads to breakthroughs and discoveries that may not have been possible otherwise. History is replete with examples of individuals and organizations that turned their failures into steppingstones for innovation, proving that setbacks can be catalysts for groundbreaking advancements.

Additionally, setbacks teach us the importance of perseverance. Success is rarely achieved overnight, and the journey is often fraught with difficulties. However, those who persist despite setbacks demonstrate an unwavering commitment to their goals. This persistence not only enhances our problem-solving skills but also reinforces our dedication, ultimately paving the way for success.

Furthermore, setbacks foster humility and empathy. Experiencing failure reminds us of our vulnerabilities and limitations, making us more compassionate towards ourselves and others. It allows us to connect with others on a deeper level, as we recognize the shared human experience of struggle and triumph. This empathy becomes a powerful tool for building meaningful relationships and fostering a supportive community.

Ultimately, the key to learning from setbacks lies in our willingness to confront them head-on. It requires a shift in mindset, from viewing failure as a source of shame to recognizing it as an integral part of the learning process. By embracing setbacks as opportunities for growth, we can transform our approach to challenges and unlock our full potential.

In the grand tapestry of life, setbacks are not blemishes but integral threads that contribute to the richness of our experiences. They shape us, refine us, and propel us towards a future where success is not just a destination but a journey marked by continuous learning and self-discovery.

Turning Mistakes into Opportunities

Mistakes are often seen as the blemishes on the canvas of our lives. They are the moments we wish to erase, the choices we wish to undo, and the instances we often want to forget. However, what if these missteps were not marks of failure but rather steppingstones to success? In shifting our perspective, we open ourselves to a world where mistakes become the catalyst for growth and innovation.

Every error we make is a lesson disguised in the cloak of failure. Instead of viewing mistakes as mere blunders, consider them as invaluable feedback. Each miscalculation is an opportunity to learn something new about ourselves, our processes, and the world around us. By analyzing what went wrong, we can identify gaps in our knowledge or skills and take actionable steps to improve.

Moreover, mistakes foster resilience. Persevering through setbacks builds mental fortitude and teaches us to adapt in the face of adversity. This resilience is crucial in both personal and professional domains, where challenges are inevitable. Those who learn to navigate their mistakes with grace and determination often emerge stronger and more capable.

Innovation, too, is born from the ashes of failure. Many of the world's greatest inventions and breakthroughs were the result of repeated trials and errors. The light bulb, the airplane, and even the internet were not instantaneous successes but rather the culmination of numerous failed attempts. These pioneers understood that each mistake brought them one step closer to their goal, and they leveraged their failures to fuel their creativity.

In a professional setting, encouraging a culture that views mistakes as opportunities rather than setbacks can drive a team towards unprecedented success. When employees feel safe to take risks without the fear of harsh repercussions, they are more likely to experiment and innovate. This environment not only fosters creativity but also enhances team morale and cohesion.

On a personal level, embracing mistakes can lead to greater self-awareness and personal growth. When we allow ourselves the grace to fail, we develop a deeper understanding of our strengths and weaknesses. This self-awareness is a powerful tool that can guide us in making more informed decisions in the future.

Furthermore, mistakes can serve as a bridge to empathy and connection. When we openly acknowledge our missteps, we create space for vulnerability, which can strengthen our relationships with others. Sharing our experiences of failure can inspire and support those around us, creating a community built on understanding and mutual encouragement.

The journey of turning mistakes into opportunities requires a mindset shift. It involves viewing each setback as a chance to learn, grow, and innovate. By adopting this perspective, we transform our relationship with failure. Instead of being paralyzed by the fear of making mistakes, we become empowered by the possibilities they present. In this way, failure ceases to be a source of shame and becomes a powerful tool for success.

Conquer your fear of failure

Failure often looms over us like a dark cloud, casting shadows on our aspirations and dreams. It whispers words of doubt and plants seeds of insecurity. Yet, to truly live without the shackles of shame, we must confront this fear head-on. The fear of failure is a formidable adversary, but it is not invincible.

Consider the moments when fear has held you back. How many opportunities were lost because the risk of failing seemed too great? How often have you chosen the familiar path, not out of desire, but out of fear of the unknown? These are the moments that define our relationship with failure. But what if we could redefine that relationship? What if failure was not something to dread but a stepping-stone to growth?

The first step in conquering this fear is understanding its nature. Fear of failure is not an innate trait, but a learned response, often rooted in past experiences or societal pressures. Reflect on where this fear originated. Was it a critical remark from a teacher, a parent, or perhaps a peer? Recognizing its source can be liberating and is the first step towards dismantling its power.

Once we understand fear's origins, we can begin to challenge its hold over us. One effective strategy is to re-frame our perception of failure. Instead of viewing it as the end, see it as a beginning. Failure is not a reflection of our worth but a testament to our courage to try. It provides invaluable lessons that success often cannot. Each setback is an opportunity to learn, adapt, and emerge stronger.

To conquer fear, we must also cultivate a mindset that embraces imperfection. The pursuit of perfection is often what fuels our fear of failure. We must learn to accept that mistakes are part of the human experience. By setting realistic expectations and allowing ourselves the grace to err, we diminish the fear that failure will define us.

Surrounding ourselves with a supportive community can also be transformative. Engage with individuals who embrace failure as part of success, who see potential in every attempt, regardless of the outcome. Their perspectives can be contagious, encouraging us to view our own failures through a more forgiving lens.

Visualization is another powerful tool in overcoming fear. Picture yourself facing a challenge and failing. Allow yourself to experience the emotions that accompany that failure. Then, visualize yourself rising from it, having learned something new. This exercise can desensitize us to the fear, showing us that failure is not the end but a part of the journey.

Ultimately, conquering the fear of failure requires action. We must take risks, step out of our comfort zones, and embrace the possibility of not succeeding. Each step we take chips away at the fear, building confidence and resilience. The more we face failure, the less power it holds over us.

In a world that often equates success with worth, choosing to confront the fear of failure is a radical act of self-liberation. It allows us to pursue our passions without the weight of shame and opens doors to opportunities that fear would have kept closed. By transforming our relationship with failure, we reclaim our power and pave the way for a life without limits.

Three

THE POWER OF VULNERABILITY

Owning Your Story

In a world where perfection is often celebrated and mistakes are hidden behind closed doors, the power of acknowledging and owning your story is revolutionary. It is a radical act of self-acceptance in a society that frequently demands conformity and flawlessness. To own your story is to declare that your experiences, both the triumphs and the failures, are valuable and integral to who you are today.

The first step in this journey is to discard the notion that failure equates to personal inadequacy. Too often, we internalize our setbacks, allowing them to define our self-worth. This mindset creates a cycle of shame that stifles growth and creativity. However, by shifting our perspective, we can begin to see failure not as a reflection of our character but as a cornerstone of personal development.

Acknowledging your past mistakes and shortcomings is not an admission of defeat; rather, it is a bold statement of resilience. It takes courage to face the uncomfortable truths about ourselves and to resist the urge to bury them beneath layers of denial. By confronting these truths, we strip them of their power to induce shame and instead transform them into lessons.

Moreover, owning your story fosters a deeper connection with others. When you share your vulnerabilities, you invite empathy and understanding. People are drawn to authenticity; they find comfort in knowing that they are not alone in their struggles. By opening up about your experiences, you create a safe space for others to do the same, cultivating a community built on genuine human connection.

It's important to recognize that no story is too insignificant to be owned. Each narrative, no matter how small or seemingly inconsequential, contributes to the tapestry of your life. By embracing every aspect of your story, you honor the complexity and diversity of your experiences. This acceptance empowers you to move forward with a sense of purpose and clarity.

Owning your story also means taking responsibility for the choices you make. While it is essential to forgive yourself for past mistakes, it is equally important to learn from them and make conscious decisions moving forward. This accountability fosters personal growth and helps you align your actions with your values.

In the realm of personal development, owning your story is a transformative act. It requires vulnerability and honesty, but the rewards are immeasurable. By accepting and embracing your entire narrative, you unlock the potential for deeper self-awareness, stronger relationships, and a life free from the constraints of shame.

Ultimately, the journey to owning your story is not about achieving perfection; it is about celebrating progress and resilience. It is about recognizing that every chapter, no matter how challenging, contributes to the richness of your life. By owning your story, you reclaim your power and pave the way for a future defined not by shame, but by strength and authenticity. This is the path to failing without shame and living a life of true fulfillment.

Connecting Through Authenticity

In a world that often values perfection and success above all else, the fear of failure can be paralyzing. We're taught to hide our flaws and mask our vulnerabilities, but what if the key to overcoming shame lies in embracing our true selves? Authenticity is not just a buzzword; it's a powerful tool that can transform our relationships, our careers, and ultimately, our lives.

Authenticity begins with self-awareness. To connect with others genuinely, we must first understand and accept who we are. This means acknowledging our strengths and weaknesses, our dreams and fears, and even our past mistakes. By confronting these aspects of ourselves, we can begin to dismantle the walls of shame that have been built around us.

When we present an authentic version of ourselves to the world, we invite others to do the same. This creates a ripple effect of openness and honesty, fostering environments where vulnerability is not seen as a weakness but as a strength. In these spaces, the fear of failure diminishes, and the stigma of shame loses its power. People feel empowered to take risks, knowing that their worth is not defined by their successes or failures but by their integrity and authenticity.

By connecting through authenticity, we build trust. In personal relationships, this means being open about our feelings, our struggles, and our needs. It means having the courage to admit when we're wrong and the humility to listen when others speak. In professional settings, authenticity can lead to more meaningful collaborations and innovations. When team members feel safe to express their true thoughts and ideas, creativity flourishes, and solutions are found more effectively.

Authenticity also has a profound impact on our mental well-being. When we stop hiding behind facades and start living in alignment with our true selves, we experience a sense of liberation. The constant pressure to maintain appearances is lifted, and we can focus on what truly matters. This newfound freedom reduces stress and anxiety, allowing us to approach challenges with a clearer mind and a more resilient spirit.

Moreover, authenticity fosters deeper connections. When we are genuine, we attract like-minded individuals who appreciate us for who we are, not for who we pretend to be. These connections are the foundation of supportive networks that can help us navigate life's ups and downs without judgment. In these communities, failure is not a source of shame but an opportunity for growth and learning.

Living authentically requires courage and commitment. It means being willing to stand out, to be different, and to face criticism. But the rewards are immeasurable. By embracing authenticity, we not only liberate ourselves from the shackles of shame but also inspire others to do the same. Together, we can create a world where failure is not feared but embraced as a natural part of the human experience, a world where our true selves are celebrated, and where shame has no place.

Strength in Openness

In a world where vulnerability is often masked by layers of self-preservation, the act of being open about one's failures can seem daunting. Yet, it is in this openness that true strength resides. Embracing the moments when our efforts fall short, instead of hiding them, allows us to grow and learn in unprecedented ways. When we acknowledge our imperfections, we invite others to do the same, creating a culture of acceptance and understanding.

The societal norm tends to glorify success and shun failure, pushing individuals to conceal their setbacks. This culture of concealment festers shame and stifles personal growth. However, choosing to reveal our failures can dismantle the stigma associated with not meeting expectations. It challenges the narrative that success is the only path worth discussing, offering a richer, more nuanced story of human endeavor.

Openness about failure is not about dwelling on the negative but about fostering resilience. By sharing our experiences, we can dissect our failures, understand their roots, and build strategies to overcome them. This process of reflection and analysis is crucial for personal development. It transforms failure from a static state to a dynamic opportunity for improvement.

Moreover, when we openly discuss our failures, we contribute to a collective wisdom that benefits everyone. Each story of failure carries lessons that others can learn from, sparing them similar pitfalls. This exchange of experiences not only enhances individual understanding but also strengthens communal bonds. The shared knowledge becomes a powerful tool for collective advancement, breaking down the barriers imposed by fear and isolation.

In professional environments, openness about failure can lead to innovation and creativity. When team members feel safe to express their mistakes without fear of judgment, they are more likely to take risks and explore unconventional solutions. This atmosphere of trust and transparency can be the breeding ground for groundbreaking ideas and improvements.

On a personal level, being open about our shortcomings fosters authenticity. It strips away the facade of perfection and allows us to present our true selves to the world. This authenticity is magnetic, drawing others who resonate with our experiences and creating deeper, more meaningful connections. It is a liberating experience to be accepted for who we are, flaws and all, and it starts with the courage to be open.

In essence, the strength in openness lies in its ability to transform shame into empowerment. It is a radical act of courage to stand amidst our failures and still hold our heads high. This act not only liberates us from the shackles of perfectionism but also inspires others to do the same. By redefining failure as an integral part of the learning process, we pave the way for a more compassionate and understanding world.

Therefore, the next time you face a setback, consider the power of openness. Share your story, not as a tale of defeat, but as a testament to resilience and growth. In doing so, you contribute to a world where failure is not feared but embraced as a steppingstone to greater achievements. Let your openness be the beacon that guides others out of the shadows of shame and into the light of understanding and strength.

The Role of Empathy

Imagine standing at a crossroads where the weight of failure looms over you like an ominous cloud. It is in this precarious moment that empathy emerges as a beacon of hope, illuminating the path toward understanding and healing. Empathy is not merely a passive emotion; it is an active, transformative force that allows us to connect with the experiences of others and, crucially, with ourselves.

When faced with the harsh reality of failure, many are quick to cast judgment, often exacerbating feelings of shame and inadequacy. This is where empathy steps in, offering a compassionate lens through which to view our setbacks. By fostering a climate of understanding, empathy encourages us to see failure not as a definitive end but as a stepping-stone toward growth and resilience. It shifts the narrative from one of self-condemnation to one of self-compassion.

Empathy's power lies in its ability to transcend the superficial layers of our experiences, delving into the deeper emotional currents that drive our actions. It invites us to pause, reflect, and truly listen—not just to the words spoken by others but to the unspoken emotions that reside within. In doing so, empathy dismantles the barriers of isolation that often accompany failure, allowing us to feel seen, heard, and valued.

Consider the profound impact empathy has in a community or organizational setting. When leaders and peers approach failures with empathy, they cultivate an environment where innovation thrives. Mistakes become opportunities for learning rather than stigmas to be avoided at all costs. This shift in perspective encourages a culture of openness and collaboration, where individuals feel safe to take risks and explore new ideas without the fear of ridicule or reprimand.

Empathy also plays a crucial role in personal relationships, where the ripple effects of failure can strain even the strongest bonds. By practicing empathy, we extend a bridge of understanding to those we care about, reassuring them that they are not alone in their struggles. This shared sense of vulnerability strengthens connections, fostering a sense of solidarity and mutual support.

Moreover, empathy is not a finite resource; it is a skill that can be nurtured and developed over time. By actively practicing empathy, we hone our ability to navigate the complexities of human emotion, enhancing our capacity for both self-awareness and emotional intelligence. This, in turn, equips us with the tools necessary to face our own failures with grace and dignity.

In the face of failure, empathy offers a powerful antidote to shame. It reminds us that we are all part of a shared human experience, bound by our imperfections and aspirations. By embracing empathy, we unlock the potential for personal transformation, viewing our failures not as scars but as badges of courage, each one a testament to our resilience and capacity for growth.

Let empathy be your guide as you traverse the challenging terrain of failure. Allow it to illuminate your path, offering a compassionate perspective that transforms shame into strength and isolation into connection. In this way, empathy becomes not only a tool for understanding but a catalyst for profound personal and collective evolution.

Four

THE INTERSECTION OF FAILURE AND MENTAL HEALTH

Negative Impacts of Failure

In a world that prizes success above all else, the concept of failure often looms as a formidable shadow. Many of us have been conditioned to view failure as a personal shortcoming, a stain on our worthiness, and a definitive end to our aspirations. This perception is not just disheartening; it's profoundly damaging. The negative impacts of failure, when left unchecked and misinterpreted, can seep into every corner of our lives, affecting our mental health, relationships, and self-esteem.

At the heart of this issue is the societal stigma attached to failure. From a young age, people are taught to equate their failures with inadequacy, a belief that can lead to a persistent fear of trying anything new. This aversion to risk stifles creativity and innovation. It chains individuals to the familiar and prevents them from reaching their full potential. The fear of failure becomes a self-imposed barrier, restricting personal growth and professional advancement.

Moreover, the psychological toll of failure can be immense. The emotional response to failure often includes feelings of shame, guilt, and embarrassment. These emotions can spiral into a cycle of self-doubt and anxiety, crippling one's ability to move forward. When people internalize their failures, they begin to question their capabilities, leading to a diminished sense of self-worth. This erosion of confidence can have long-lasting effects, making it difficult to recover and pursue new opportunities.

The social implications of failure are equally profound. In a culture that celebrates only success, those who fail often find themselves isolated. The fear of judgment from peers and the community can lead individuals to withdraw, cutting themselves off from valuable support networks. This isolation exacerbates the emotional strain of failure, creating a sense of loneliness and alienation. The stigmatization of failure not only affects the individual but also discourages open dialogue about setbacks, perpetuating a cycle of silence and shame.

Furthermore, the professional consequences of failure can be daunting. In the workplace, a single failure can overshadow a history of accomplishments, leading to missed promotions and opportunities. This can foster a toxic environment where employees are afraid to take risks or propose innovative ideas. The fear of making mistakes stifles the potential for growth and progress, both for individuals and organizations.

Financially, the burden of failure can be overwhelming. For entrepreneurs and business owners, a failed venture can result in significant financial loss, impacting not just their livelihood but also their confidence to try again. The economic ramifications extend beyond personal losses, affecting families and communities dependent on these ventures.

Yet, the most insidious impact of failure is the way it shapes our identity. When failure is perceived as a defining trait rather than a temporary setback, it can lead to a fixed mindset, where individuals see their abilities as static and unchangeable. This mindset limits resilience and adaptability, essential traits for overcoming challenges and achieving long-term success.

To navigate the negative impacts of failure, it is crucial to redefine our understanding of what it means to fail. By shifting the narrative from one of shame and inadequacy to one of learning and growth, we can begin to dismantle the barriers that hold us back. In doing so, we can foster a culture that values resilience, encourages innovation, and embraces the opportunities that come with every setback.

Finding Strength in Failure

In the grand tapestry of human endeavors, failure is often seen as an unsightly thread—something to be hidden or cut away. Society tends to celebrate success and shun failure, leaving many to carry the burden of shame when they fall short of their goals. However, what if failure is not the villain it is often portrayed to be? Instead, it is an essential part of the journey, a catalyst for growth, and a steppingstone towards eventual success.

Consider this: every groundbreaking invention, every monumental achievement, has been preceded by countless failed attempts. Thomas Edison, whose name is synonymous with innovation, famously stated, "I have not failed. I've just found 10,000 ways that won't work." This mindset reframes failure not as a dead-end but as a valuable part of the learning process. It is through these missteps that we gain the insight necessary to refine our approach and ultimately reach our goals.

Failure offers the opportunity to develop resilience. Each setback is a chance to reflect, reassess, and re-calibrate. It teaches us to persevere in the face of adversity and to adapt to changing circumstances. This resilience is a powerful trait, one that can be applied to all areas of life, from personal relationships to professional endeavors. Those who learn to harness the lessons of failure are often the ones who emerge stronger and more prepared for future challenges.

Moreover, failure strips away the illusion of perfection, allowing us to connect more authentically with others. When we share our failures, we open the door to vulnerability and honesty. This transparency fosters deeper relationships, as others see that they are not alone in their struggles. In a world that often values appearances over substance, embracing failure allows us to prioritize authenticity and genuine connection.

Failure also encourages innovation and creativity. When the fear of failure is removed, individuals are more likely to take risks and explore new ideas. This willingness to experiment can lead to breakthroughs that might otherwise remain undiscovered. By viewing failure as a natural part of the creative process, we can foster an environment where innovation thrives.

Furthermore, failure can be a powerful motivator. The sting of falling short can ignite a renewed determination to succeed. It serves as a reminder of what is at stake and what can be achieved with persistence and dedication. This drive can push us to work harder, think deeper, and strive for excellence in ways we might not have considered before.

The key to finding strength in failure lies in our perspective. If we view it as a teacher rather than a tyrant, we can unlock its potential to propel us forward. By embracing failure as an integral part of the human experience, we can shed the shame that often accompanies it and instead focus on the growth and opportunity it provides.

Ultimately, to fail is not to fall, but to learn. It is an invitation to rise stronger, armed with the knowledge and insight gained from our experiences. In embracing this truth, we find the courage to pursue our dreams without fear, knowing that even in failure, there is strength.

Coping Strategies

Imagine a world where failure is not a source of shame but a steppinstone to success. A place where each stumble is an opportunity to grow, learn, and ultimately rise stronger. This vision is not only possible but within reach, and the key lies in the strategies we employ to cope with failure.

The first step in transforming failure from a source of embarrassment to a catalyst for growth is changing our mindset. The power of perception is immense. By viewing failures as valuable lessons rather than personal shortcomings, we open the door to resilience and innovation. This shift in perspective allows us to dissect the experience, understand its nuances, and apply the insights gained to future endeavors.

Acknowledging emotions is crucial in this process. Suppressing feelings of disappointment, anger, or frustration only prolongs their impact. Instead, embracing these emotions and allowing oneself to process them can lead to healthier mental states and clearer thinking. Journaling or speaking with a trusted confidant can be powerful tools in this emotional unpacking, providing clarity and new perspectives.

Another vital strategy is setting realistic expectations. Often, the fear of failure stems from unrealistic standards we impose on ourselves. By breaking down goals into manageable steps and celebrating small victories along the way, we can alleviate the pressure of perfection. This approach not only makes the journey more enjoyable but also reinforces the notion that progress, not perfection, is the true marker of success.

Developing a robust support system is equally important. Surrounding oneself with individuals who understand and empathize with the challenges of failure can provide a safety net of encouragement and advice. Whether it's a mentor, a peer group, or a community of like-minded individuals, having a network to lean on during tough times can make all the difference.

Furthermore, cultivating a habit of self-compassion is essential. Treating oneself with the same kindness and understanding one would offer a friend in similar circumstances helps mitigate the harsh inner critic that often accompanies failure. Practicing mindfulness can aid in fostering this self-compassion, allowing for a more balanced and forgiving approach to personal setbacks.

Learning to pivot is another powerful coping strategy. When faced with a failed endeavor, the ability to adapt and explore alternative paths is crucial. This flexibility not only increases the likelihood of eventual success but also builds resilience and fosters creativity. Embracing a mindset of adaptability turns obstacles into opportunities for innovation.

Finally, it's important to remember that failure is a universal experience. It is not an indictment of one's abilities or worth but a shared aspect of the human condition. By normalizing conversations around failure and sharing experiences openly, we contribute to a culture that values growth over perfection.

In navigating the landscape of failure, these strategies serve as invaluable tools. They empower us to face setbacks with courage and transform moments of shame into steppinstones toward personal and professional fulfillment. Through these practices, we can redefine failure not as a terminal point but as a necessary part of the journey toward success.

Seeking Support

In the midst of grappling with failure, the instinct to retreat into solitude often feels overwhelming. We build walls, seeking refuge in our own thoughts, believing it to be a fortress of safety. Yet, in these moments of perceived isolation, the power of seeking support becomes paramount.

The societal narrative often paints seeking help as a sign of weakness, a notion that further entrenches the stigma surrounding failure. However, reframing this perspective reveals a profound truth: reaching out is a testament to strength and resilience. It is an acknowledgment of our humanity, a recognition that we are inherently social beings who thrive in connection.

Imagine a world where individuals openly share their setbacks, where vulnerability is met with empathy rather than judgment. This vision is not a distant utopia but a tangible reality we can cultivate. By seeking support, we dismantle the myth that failure is an anomaly to be hidden. Instead, it becomes a shared experience, a collective narrative that fosters growth and understanding.

Consider the power of community. When we reach out, we tap into a reservoir of experiences and wisdom. Those who have traversed similar paths offer invaluable insights, providing perspectives that illuminate new possibilities. They become mirrors reflecting our potential, reminding us that failure is not a dead end but a detour on the road to success.

Furthermore, seeking support nurtures our well-being. The emotional toll of failure, when borne alone, can be debilitating. It manifests in self-doubt, anxiety, and a pervasive sense of inadequacy. Yet, in the company of others, these burdens are lightened. Conversations become cathartic, allowing us to process emotions and gain clarity. Support networks act as a safety net, catching us when we stumble, offering reassurance that we are not alone in our struggles.

Moreover, reaching out cultivates a cycle of reciprocity. By seeking support, we open ourselves to receiving, and in turn, become more attuned to the needs of others. This exchange fosters a culture of empathy and solidarity, where the success of one becomes the success of many. It is a reminder that in our interconnectedness, we find strength.

The journey of seeking support is not without its challenges. It requires vulnerability, a willingness to expose our imperfections. Yet, it is in this vulnerability that we find liberation. By dismantling the facade of invincibility, we allow others to see us in our entirety, fostering authentic connections that transcend superficial interactions.

As you navigate the complexities of failure, consider the transformative impact of seeking support. It is an act of courage, a declaration that you are worthy of understanding and compassion. In reaching out, you not only empower yourself but also contribute to a paradigm shift, where failure is embraced as a natural and necessary part of the human experience.

In this collective journey, we find the strength to rise, to learn, and to grow. Seeking support is not a sign of defeat but a powerful step towards reclaiming our narrative, transforming shame into resilience, and failure into an opportunity for profound personal evolution.

Cultivating Self-compassion

In a world that often glorifies perfection and stigmatizes failure, the internal dialogue we maintain can be our greatest ally or our harshest critic. Imagine a friend who, in times of your distress, chooses to berate you rather than offer support. Would you not question the sincerity of such a friendship? Yet, how often do we find ourselves being that unfriendly companion to our own selves? The path to overcoming the crippling effects of shame lies in the ability to cultivate self-compassion—a powerful tool that allows us to fail without fear.

Self-compassion is the practice of treating oneself with the same kindness and understanding that we would extend to a dear friend. It is about recognizing that imperfection is part of the human experience, and it is this very imperfection that connects us all. When we falter, it is crucial to respond with empathy rather than self-criticism. This gentle approach fosters a nurturing environment within us, where growth is possible and shame loses its grip.

The misconception that self-compassion equates to self-indulgence or lack of accountability is a barrier that must be dismantled. Self-compassion does not mean letting ourselves off the hook; rather, it involves taking responsibility with an attitude of forgiveness and encouragement. By accepting our mistakes as opportunities for learning, we transform shame into a steppingstone for personal development.

Consider the voice of self-criticism, which often speaks with the harshness of an unyielding judge. It tells us that we are not good enough, that our failures define us. In contrast, the voice of self-compassion is that of a wise mentor, whispering that mistakes are but a small part of our journey, not the entirety of our story. It reminds us that our value is inherent, not contingent on our successes or failures.

The practice of self-compassion begins with mindfulness. By acknowledging our feelings without judgment, we create space for understanding and healing. This mindfulness allows us to observe our thoughts and emotions, recognizing them as transient rather than definitive. It is in this awareness that we find the strength to challenge negative self-talk and replace it with affirmations of worthiness and resilience.

Moreover, self-compassion fosters a sense of connectedness. When we accept our imperfections, we become more empathetic towards others, recognizing that everyone struggles and falters. This shared humanity diminishes isolation, allowing us to offer and receive support without the fear of judgment. In this way, self-compassion not only liberates the self but also enriches our relationships with others.

To fail without shame is to embrace self-compassion as a guiding principle. It is an invitation to be gentle with ourselves, to practice patience and understanding in the face of adversity. By cultivating self-compassion, we dismantle the barriers of shame, paving the way for authentic growth and fulfillment. Let us choose to be our own allies, nurturing our inner landscape with kindness and compassion, and watch as shame loses its power over us. In this compassionate embrace, we find the courage to not only confront failure but to thrive in its presence.

Five

REFRAMING FAILURE IN SOCIETY

Cultural Perceptions of Failure

In the grand tapestry of human history, failure has often been painted with a brush of misfortune and disgrace. Yet, what if this perception is not a universal truth but rather a cultural construct? Across the globe, societies have woven intricate narratives around the concept of failure, shaping how individuals perceive setbacks and, in turn, how they respond to them.

Consider the Western ideals prevalent in regions like the United States and much of Europe, where the ethos of individualism reigns supreme. Here, success is often measured by personal achievement and the accumulation of accolades. Failure, therefore, becomes a deeply personal experience, one that is often internalized as a reflection of one's abilities or worth. This perception can lead to a debilitating fear of failure, where the mere possibility of a setback is enough to deter risk-taking and innovation. The cultural narrative suggests that to fail is to falter in one's quest for self-ful-fillment, which can foster an environment where failure is shunned and hidden away.

Contrast this with many Eastern cultures, where the concept of collectivism is more pronounced. In places like Japan or China, failure is often viewed through the lens of community and family. Here, an individual's setback can be perceived as a reflection on their family or group, inter-twining personal failure with communal shame. This can create a pressure to succeed not just for oneself, but for the collective, leading to a different kind of anxiety around failure. Yet, within these same cultures, there often exists a paradoxical appreciation for perseverance and resilience. The Japanese concept of "Nana korobi ya oki"—fall down seven times, stand up eight—epitomizes the idea that failure is an expected part of life's journey, and the true measure of a person is their ability to rise again.

Meanwhile, in some African and Indigenous cultures, failure is perceived more as a natural part of learning and growth. The emphasis is placed on the lessons learned and the wisdom gained from each experience, rather than the shame of the setback itself. This perspective fosters a more forgiving environment, where failure is not stigmatized but rather seen as an opportunity for communal learning and growth. By focusing on the collective journey rather than individual achievement, these cultures often provide a supportive backdrop for individuals to explore and innovate without the paralyzing fear of failure.

These varied cultural perceptions of failure highlight an essential truth: how we view setbacks is deeply influenced by the narratives we are told and the values we hold dear. It is not failure itself that is inherently shameful, but rather the meaning we ascribe to it. By understanding these cultural dimensions, we can begin to unravel the stigma surrounding failure, transforming it into a steppingstone rather than a stumbling block. In doing so, we open the door to a world where failure is not feared but embraced as a vital part of the human experience. Through this lens, we can foster cultures that encourage innovation, resilience, and ultimately, a more profound understanding of success itself.

In the quest to redefine failure, it is crucial to recognize and challenge these cultural narratives, paving the way for a more inclusive, empathetic understanding of what it truly means to fail without shame. This is not merely a shift in perspective, but a call to action—a revolution in how we perceive and respond to the inevitable setbacks that life presents. By doing so, we can cultivate a society that celebrates growth, learning, and the indomitable spirit of the human endeavor.

Media's Role in Shaping Views

In the modern age, media is an omnipresent force, wielding immense power in shaping societal views and individual perceptions. It is an intricate tapestry of narratives spun by diverse channels—news outlets, social media platforms, television, and films—each contributing to a collective consciousness that often dictates societal norms and personal beliefs. This chapter delves into the profound influence media exerts on our perception of success and failure, urging readers to critically assess and challenge these pervasive narratives.

The media often presents a skewed portrayal of success, glorifying achievements while glossing over the inevitable struggles and failures that accompany any worthwhile endeavor. Celebrities, entrepreneurs, and public figures are frequently depicted as paragons of success, with their journeys seemingly devoid of setbacks. This curated image fosters unrealistic expectations, leading individuals to equate success with perfection and view failure as a shameful anomaly. It is imperative to recognize that these portrayals are far from the comprehensive truth.

Consider the stories of renowned innovators and leaders, often heralded as overnight successes. The reality behind these tales is a series of trials, errors, and learning experiences that are conveniently omitted from mainstream narratives. By emphasizing only the triumphs, media narratives can inadvertently instill a fear of failure, discouraging individuals from taking risks or pursuing unconventional paths. This fear becomes a barrier, stifling creativity and innovation.

Moreover, media platforms thrive on sensationalism, often amplifying stories of failure to attract attention. Headlines that emphasize downfall and disgrace contribute to a culture that stigmatizes failure. This portrayal not only affects public perception but also influences how individuals perceive their own setbacks. The shame associated with failure, perpetuated by media narratives, can lead to self-doubt and hinder personal growth.

However, media also holds the potential to reshape these narratives and promote a more balanced understanding of success and failure. By highlighting stories of resilience, perseverance, and eventual triumphs over adversity, media can inspire individuals to view failure as a natural and necessary part of the journey towards success. These stories can empower people to embrace failure as a learning opportunity, fostering a mindset that values growth over perfection.

The responsibility lies not only with media creators but also with consumers. Media literacy is crucial in discerning between factual representations and sensationalized narratives. By critically engaging with media content, individuals can challenge the prevailing notions of success and failure. They can seek out diverse perspectives and stories that reflect the multifaceted nature of human experience.

In a world dominated by media, it is essential to cultivate a discerning mindset, one that questions the narratives presented and seeks a deeper understanding of the complexities of success and failure. By doing so, individuals can liberate themselves from the shackles of shame associated with failure and embrace a more authentic and empowering view of their journeys. The media's role is pivotal, but it is ultimately up to each person to navigate these narratives with a critical eye and an open mind, transforming failure from a source of shame into a catalyst for growth.

Education Systems and Failure

In the pursuit of academic excellence, our education systems have long been heralded as the bedrock of societal progress. Yet, beneath the polished veneer of success lies an uncomfortable truth: these systems often cultivate a profound fear of failure. This fear stifles creativity, discourages risk-taking, and ultimately undermines the very purpose of education itself.

Education should be a fertile ground for exploration and innovation, a space where students are encouraged to push boundaries and venture beyond the familiar. Yet, the reality is starkly different. The rigid structures and standardized assessments that dominate our schools prioritize conformity over creativity, rewarding students who excel at rote memorization while marginalizing those who dare to think differently. In this environment, failure becomes a stigma, a mark of inadequacy rather than a steppingstone to growth.

Consider the impact on the individual student. From an early age, children are conditioned to equate success with high grades and accolades. This conditioning fosters an aversion to failure, creating a mindset where mistakes are to be avoided at all costs. Consequently, students become risk-averse, opting for safe choices that guarantee success rather than daring to explore uncharted territories. This fear of failure not only hinders personal development but also stifles the innovative potential that lies dormant within each learner.

Furthermore, this systemic aversion to failure has broader societal implications. By teaching students to fear failure, education systems inadvertently limit their ability to adapt to an ever-changing world. In the face of global challenges that demand creative solutions, the inability to embrace failure as a natural part of the learning process becomes a significant handicap. It is through failure that we learn resilience, develop problem-solving skills, and cultivate the ability to think critically. Without these skills, students are ill-equipped to navigate the complexities of modern life.

Yet, the tide is turning. There is a growing recognition that failure is not the antithesis of success but an integral part of it. Progressive education models are emerging, advocating for a shift in perspective where failure is not demonized but celebrated as a learning opportunity. These models emphasize experiential learning, encouraging students to engage in projects that challenge their assumptions and push the boundaries of their understanding. By reframing failure as a valuable component of the learning process, these approaches foster an environment where students feel empowered to take risks, innovate, and grow.

However, transforming education systems to embrace failure requires a concerted effort from educators, policymakers, and society at large. It demands a cultural shift that redefines success, moving away from narrow metrics and towards a more holistic understanding of achievement.

We must cultivate educational environments that nurture curiosity, encourage experimentation, and celebrate the unique journey of each learner.

In doing so, we will not only prepare students for the challenges of tomorrow but also unlock their potential to create a world where failure is not a source of shame but a catalyst for innovation and growth. By redefining our relationship with failure, we can build education systems that truly empower individuals to become resilient, adaptable, and capable of shaping a brighter future for all.

Creating Supportive Environments

Imagine a world where failure is not a mark of disgrace but a steppingstone to greatness. This vision can only be realized when we foster environments that nourish growth and learning. In our pursuit of creating spaces where failure is not feared but embraced, we must first dismantle the barriers that breed shame and isolation.

Too often, individuals are conditioned to view setbacks as personal shortcomings, leading to self-doubt and a reluctance to take risks. This mindset stifles creativity and innovation, halting progress in its tracks. To counteract this, we must cultivate environments that view failure as an integral part of the learning process. By doing so, we empower individuals to explore new ideas and push boundaries without the fear of judgment or ridicule.

A supportive environment begins with open communication. Encouraging dialogue about failures and successes alike creates a culture of transparency. When people feel safe to share their experiences, they are more likely to seek feedback and learn from others. This exchange of ideas not only fosters personal growth but also strengthens the community as a whole.

Moreover, mentorship plays a pivotal role in creating supportive environments. Mentors who share their own stories of failure and resilience provide invaluable lessons and reassurance that setbacks are not the end but merely a part of the journey. Through guidance and encouragement, mentors help individuals navigate challenges and develop the resilience needed to persevere.

Additionally, recognition of effort rather than just outcomes is crucial in nurturing a supportive space. Celebrating the process and the courage it takes to attempt something new reinforces the idea that failure is not a dead end but a valuable learning opportunity. This shift in focus from results to effort encourages risk-taking and innovation, essential components of growth.

Creating supportive environments also involves addressing systemic issues that perpetuate the stigma of failure. Institutions and organizations must reevaluate their policies and practices to ensure they do not inadvertently penalize failure. Instead, they should implement systems that reward experimentation and learning from mistakes. This requires a cultural shift that values adaptability and resilience over perfection.

Furthermore, the role of leadership cannot be understated. Leaders set the tone for the environment, and their attitudes toward failure can significantly impact the organization's culture. Leaders who model vulnerability and resilience inspire their teams to do the same. By demonstrating that failure is a natural part of the process, leaders can cultivate a culture where innovation thrives.

In conclusion, the journey to creating supportive environments is not without its challenges, but the rewards are immeasurable. By fostering spaces where failure is not feared but embraced, we unlock the potential for creativity, innovation, and growth. In these environments, individuals are empowered to take risks, learn from their mistakes, and ultimately achieve greater success. Together, we can build a world where failure is not a source of shame but a catalyst for progress.

Six

THE ROLE OF FAILURE IN INNOVATION

Fail Fast, Learn Fast

In a world where success is often measured by flawless execution and unbroken paths, the fear of failure looms large. Yet, within the shadows of perceived defeat lies a potent truth: the faster we allow ourselves to fail, the quicker we can rise with newfound wisdom. This perspective is not merely a strategy—it's a mindset shift, a pivotal transformation in how we perceive the hurdles on our path.

The concept of failing fast is not about courting failure for its own sake; instead, it's about accelerating the process of trial and error to uncover the most effective solutions. By embracing this approach, we liberate ourselves from the paralysis of perfectionism, which so often stalls progress and stifles innovation. When we remove the stigma associated with failure, we grant ourselves the freedom to experiment, to push boundaries, and to explore uncharted territories without the fear of judgment or ridicule.

Consider the landscape of innovation that has been shaped by those who dared to fail fast. From the tech giants who iterate their products rapidly to the artists who redefine genres through bold experimentation, the common thread is a willingness to learn from missteps. Each iteration, each failed attempt, is a steppingstone towards mastery and achievement. The key lies in viewing each failure not as an endpoint but as a vital piece of the larger puzzle.

This approach demands a shift in how we define success. Instead of seeing failure as a blemish on our record, we must recognize it as an essential component of growth. Each failure provides invaluable insights, teaching us what doesn't work and guiding us towards what might. The lessons gleaned from these experiences are often the catalysts for breakthroughs that might otherwise remain elusive.

By failing fast, we also cultivate resilience—a crucial trait in any endeavor. Resilience empowers us to bounce back, to adapt, and to persevere in the face of adversity. It's the armor that shields us from the sting of setbacks, allowing us to maintain momentum and keep our eyes on the ultimate goal. This resilience, born from the willingness to fail, is what separates those who achieve greatness from those who remain stagnant.

Furthermore, this mindset fosters a culture of continuous learning. When failure is destigmatized, individuals and teams are more likely to share their experiences openly, creating an environment where collective knowledge flourishes. Collaboration becomes more dynamic, as each member contributes to a tapestry of shared understanding, woven from the threads of individual trials.

In the grand tapestry of life, failure is not a solitary stain but a vivid hue that adds depth and richness to our journey. By failing fast, we open ourselves to a world of possibilities, where each misstep is a steppingstone, leading us closer to our aspirations. It is through this lens that we can navigate the complexities of our pursuits with courage and conviction, unburdened by the fear of falling short.

Let us redefine our relationship with failure and embrace the power of failing fast. For in doing so, we unlock the potential to learn swiftly, adapt readily, and achieve profoundly—without shame.

Creativity Born from Mistakes

Imagine a world where every error is seen not as a setback but as a wellspring of possibility. This is the transformative power of embracing mistakes, an approach that turns failures into the fertile ground from which creativity blossoms. In the tapestry of human achievement, some of the most brilliant innovations have emerged from the seemingly dark corners of error.

Consider the tale of the accidental inventor. Picture a scientist in a lab, a musician in a studio, or an artist at the easel. Each one, at some point, has faltered. Yet, it is precisely this stumble that has propelled them forward, catalyzing unexpected breakthroughs. When we cease to view mistakes as the end of the road and instead recognize them as detours leading to uncharted territories, we unlock the doors to innovation.

The key lies in the mindset shift. Mistakes should not be stigmatized as failures but celebrated as opportunities for exploration. When we allow ourselves to experiment without the fear of imperfection, we create an environment ripe for discovery. This is the crucible in which creativity is forged. The fear of making mistakes stifles creativity, while a willingness to engage with them nurtures it.

The process of trial and error is not merely a method of elimination but a dynamic dialogue with our own limitations and assumptions. It is through this dialogue that we refine our ideas, test our hypotheses, and push the boundaries of what is possible. By acknowledging that mistakes are an integral part of the creative process, we free ourselves from the confines of linear thinking and allow our imaginations to soar.

Moreover, mistakes serve as invaluable teachers. They illuminate paths we might never have considered, offering insights that can transform our understanding and approach. When we analyze our failures with an open mind, we gain wisdom that is often more profound than any success could provide. This wisdom becomes the bedrock upon which true innovation is built.

In the realm of creativity, mistakes are the catalysts that spark the flame of ingenuity. They challenge us to think differently, to question the status quo, and to explore the "what ifs" that lie beyond conventional wisdom. Each misstep is a brushstroke on the canvas of creativity, adding depth and texture to our work.

As we navigate the journey of life, we must cultivate the courage to fail. It is in this courage that we find the freedom to create without boundaries, to imagine without limits, and to innovate without fear. When we shift our perspective on mistakes, we redefine them not as blemishes on our record but as badges of honor on our creative journey.

In embracing our errors, we unlock the true potential of our creativity. We embark on a path where mistakes are not just tolerated but celebrated, where every failure is a steppingstone to greatness. Let us redefine the narrative, transforming mistakes into masterpieces and failures into triumphs. In doing so, we not only reshape our own destinies but also contribute to a world where creativity knows no bounds. This is the art of failing without shame, and it is the birthplace of all true innovation.

Risk-Taking as a Catalyst

In the annals of success stories, one element emerges consistently as a defining factor: the willingness to take risks. Often misunderstood or viewed with trepidation, risk-taking is not merely a reckless leap into the unknown; it is a calculated decision to step beyond the comfort zone and seize opportunities that others may overlook. This action, driven by a blend of courage and foresight, can be the very catalyst that transforms failure into triumph.

Consider the landscape of innovation and progress. The pioneers who have reshaped industries, disrupted markets, and altered the course of history have all shared a common trait—an audacious spirit willing to confront uncertainty. When viewed through this lens, risk-taking is not an option but a necessity. It propels individuals and organizations towards growth, challenging the status quo and fostering an environment where creativity can flourish.

The fear of failure often shackles potential, creating a barrier to progress. Yet, it's precisely this fear that can be harnessed as a powerful motivator. By redefining failure as a steppingstone rather than an endpoint, risk-taking becomes a strategic tool rather than a perilous gamble. The key lies in understanding that each risk carries lessons, each setback offers insights, and every failure is a precursor to success.

In the realm of personal development, the impact of risk-taking is equally profound. It cultivates resilience and adaptability, qualities that are indispensable in navigating life's unpredictable journey. Those who dare to take risks often find themselves developing a deeper understanding of their strengths and weaknesses, leading to heightened self-awareness and personal growth. This process of self-discovery, fueled by risk, is essential for building the confidence needed to tackle future challenges.

Moreover, risk-taking fosters a culture of innovation within organizations. Companies that encourage risk-taking among their employees often witness increased creativity and problem-solving capabilities. When individuals feel empowered to propose bold ideas without the fear of retribution, they contribute more effectively to the organization's success. This environment not only attracts top talent but also retains it, as employees feel valued and motivated to excel.

The economic benefits of risk-taking are substantial. Entrepreneurs who embrace risk are the driving force behind new ventures and startups, creating jobs and stimulating economic growth. By venturing into uncharted territories, they bring forth novel solutions to pressing challenges, paving the way for economic resilience and prosperity.

Yet, it is crucial to approach risk-taking with prudence. The most successful risk-takers employ thorough research, strategic planning, and contingency measures. They balance boldness with caution, ensuring that risks are calculated and informed. This approach minimizes potential downsides while maximizing the likelihood of favorable outcomes.

Ultimately, risk-taking is more than a mere act of bravery; it is an essential component of success. By embracing risk as a catalyst, individuals and organizations can unlock their full potential, turning aspirations into achievements. It is not the absence of fear that defines a successful risk-taker, but the determination to move forward despite it. In doing so, they redefine failure, not as a source of shame, but as a crucial part of the journey toward success.

Fostering a Culture for Innovation

Creating an environment where innovation thrives is crucial for any organization seeking to stay ahead of the curve. It's the fertile soil in which groundbreaking ideas take root and flourish. To build such a culture, we must first dismantle the traditional barriers that stifle creativity and replace them with a framework that nurtures experimentation and embraces the lessons learned from failure.

One of the fundamental steps in fostering innovation is encouraging open communication. When team members feel free to express their ideas without fear of ridicule or dismissal, they are more likely to contribute creative solutions. Leaders should promote a culture where questions are welcomed, and curiosity is rewarded. This openness leads to a flow of ideas that can ignite the spark of innovation.

Another critical component is creating a safe space for failure. In many organizations, the fear of failure can be paralyzing, stifling creativity and innovation. By redefining failure as a steppingstone to success rather than an endpoint, we can empower individuals to take calculated risks. Celebrating the learning that comes from failed attempts can transform fear into enthusiasm, driving a culture that values growth and resilience.

Moreover, diversity should be at the heart of any innovative culture. Diverse teams bring a variety of perspectives, experiences, and ideas, which can lead to more creative solutions. By fostering an inclusive environment that values every voice, organizations can tap into a wealth of untapped potential. This diversity of thought is a powerful catalyst for innovation.

Equally important is providing the resources and time necessary for innovation to occur. Allocating dedicated time for brainstorming sessions, hackathons, or skunkworks projects can encourage employees to think outside the box. When individuals are given the space to explore their ideas without the pressure of immediate results, they can experiment and iterate, leading to breakthroughs.

Incentivizing innovation is also a powerful way to cultivate a culture that supports creative thinking. Recognizing and rewarding innovative efforts, whether successful or not, reinforces the importance of taking initiative and thinking creatively. Incentives don't always have to be monetary; they can be as simple as public acknowledgment, opportunities for professional development, or additional resources to pursue promising ideas.

Leadership plays a pivotal role in shaping an innovative culture. Leaders who model curiosity, demonstrate a willingness to take risks, and show resilience in the face of setbacks inspire their teams to do the same. By setting the tone from the top, leaders can create an environment where innovation is not just encouraged but expected.

To truly embed innovation into the fabric of an organization, it must be seen as a continuous process rather than a one-time initiative. Regularly revisiting and refining processes, celebrating successes, and learning from failures should be part of the organizational rhythm. This ongoing commitment ensures that innovation becomes an integral part of the organization's DNA.

In conclusion, fostering a culture of innovation is about creating an environment where creativity is nurtured, risks are embraced, and learning is celebrated. By breaking down barriers and building a supportive framework, organizations can unlock the full potential of their teams, leading to transformative growth and success. It's not just about surviving in a competitive landscape; it's about thriving and leading the way forward.

From Failure to Success

Success is not merely the absence of failure; it is often the profound mastery of it. When confronted with the specter of failure, many see a door closing, a path ending. Yet, what if this very failure is the precursor to the most transformative success? To truly understand the alchemy of turning failure into success, one must first recognize that failure is not a dead-end but a crucial part of the success equation.

Imagine a sculptor chiseling away at a seemingly unyielding block of marble. Each strike may seem insignificant or even misguided, yet every chip falls away to reveal the masterpiece within. So it is with failure. Each setback, each misstep, is not a judgment of your capabilities but rather an integral part of the sculpting process towards ultimate success.

Consider the stories of those who have stood at the precipice of failure, only to rise and redefine their destinies.

These individuals did not allow failure to define them; instead, they used it as a steppingstone. They learned, adapted, and grew stronger. Their resilience was not born of innate talent but forged in the fires of repeated attempts and relentless perseverance.

The first step in transforming failure into success is reframing your perspective. Failure is not a reflection of your inadequacy but an opportunity for growth. It is a teacher, albeit a harsh one, that offers invaluable lessons. This shift in mindset is crucial. When you view failure through the lens of opportunity, the fear that often accompanies it dissipates, replaced by curiosity and determination.

Next, it is imperative to analyze and extract lessons from every failure. What went wrong? Why did it happen? How can it be prevented in the future? By dissecting failure with an analytical eye, you can uncover patterns and insights that illuminate the path forward. This process is akin to mining for gold within the rubble; it requires patience and diligence but yields treasures that pave the way for future success.

Moreover, resilience must become a cornerstone of your approach. Success is rarely instantaneous. It demands a tenacity that is willing to withstand the buffets of repeated failures. Cultivate a mindset that views each failure not as a defeat but as a challenge to be overcome. This unyielding spirit is the hallmark of those who have turned failure into triumph.

Lastly, foster a community that supports and encourages growth through failure. Surround yourself with those who understand the importance of perseverance and who can offer guidance and encouragement when the going gets tough. This network of support can be a lifeline, providing perspective and motivation when self-doubt threatens to derail your progress.

In the grand tapestry of life, failure is but one thread, albeit a vital one. It is interwoven with success, creating a rich and textured narrative that is uniquely yours. Embrace failure, for within its depths lies the potential for greatness. When you understand and harness the power of failure, you transform it from a stumbling block into a steppingstone towards success. The journey from failure to success is not a path to be feared but an adventure to be embraced, with each failure bringing you one step closer to the success you seek.

Seven

TRANSFORMING PERSONAL RELATIONSHIPS

Communication and Understanding

I magine a world where every conversation you have is riddled with misunderstandings and misinterpretations, where every attempt to share your thoughts and feelings is met with confusion or dismissal. This is the reality for many who struggle with effective communication, a skill that holds the key to building bridges between people, fostering empathy, and ultimately, living a life unfettered by shame and regret.

The power of communication lies not just in the words we choose but in the intent behind them. When we communicate effectively, we open doors to understanding that transcend cultural, social, and emotional barriers. It's not merely about expressing our needs or desires; it's about creating a space where others feel heard, valued, and respected. In such environments, shame loses its grip, allowing us to engage authentically and without fear of judgment.

Communication, at its core, is an art form that requires practice and patience. It is a dance of listening and speaking, of giving and receiving. When we master this art, we dismantle the walls of shame that often prevent us from reaching our full potential. Through understanding, we can transform failure into a powerful tool for growth, seeing it not as a reflection of our inadequacies but as a steppingstone towards greater self-awareness.

Understanding is the twin sister of communication. It demands that we not only hear the words spoken but also grasp the emotions and intentions behind them. When we cultivate understanding, we foster an environment where mistakes are not seen as failures but as opportunities to learn and improve. In this way, we build resilience against the shame that often accompanies failure.

In the world of communication, empathy is our greatest ally. It allows us to step into the shoes of others, to feel their joy and their pain, and to respond with compassion rather than judgment. Empathy bridges the gap between misunderstanding and understanding, transforming potential conflicts into collaborative dialogues. When we approach interactions with empathy, we create a culture of acceptance and support, where shame has no place to thrive.

Moreover, the courage to communicate openly and honestly is a testament to our willingness to face vulnerability.

It requires us to confront our fears and insecurities, to admit our mistakes, and to ask for help when needed. In doing so, we not only break free from the chains of shame but also inspire others to do the same. Our openness becomes a beacon of hope and a catalyst for change, encouraging a shift towards a more understanding and compassionate society.

It is within our grasp to redefine failure by reshaping our communication and understanding practices. Let us embrace the power of words, not as weapons of shame, but as tools of connection and growth. Let us strive to listen more deeply, speak more thoughtfully, and understand more fully, so that we may forge a path where failure is not a source of shame but a cornerstone of our shared humanity. Together, through improved communication and deeper understanding, we can learn to fail without shame and live with greater purpose and joy.

Navigating Conflicts with Grace

Conflicts, an inevitable aspect of human interaction, often evoke feelings of shame and inadequacy. However, they also present unique opportunities for personal growth and deeper understanding. The key lies in approaching these situations with grace—a quality that transforms potential chaos into constructive dialogue.

To navigate conflicts gracefully, one must first acknowledge the emotions involved. Often, the fear of confrontation stems from the potential exposure of our vulnerabilities. Yet, embracing these emotions rather than shunning them can lead to a more authentic and honest exchange. By recognizing our own feelings and those of others, we pave the way for empathy—a cornerstone in resolving disagreements.

Empathy, however, is not synonymous with agreement. Rather, it is about understanding the perspective of the other party. Listening actively and attentively to their concerns without immediately formulating a rebuttal is essential. This practice not only defuses tension but also fosters an environment where all parties feel heard and valued.

Furthermore, clear communication is paramount. Misunderstandings often arise from assumptions and unspoken expectations. Articulating one's thoughts and feelings with clarity and respect can prevent conflicts from escalating. This involves expressing oneself honestly while remaining open to feedback. By maintaining a calm and composed demeanor, even in the face of disagreement, we demonstrate respect for both ourselves and the other person.

Another aspect of graceful conflict navigation is the ability to compromise. It is crucial to recognize that resolution does not always mean victory for one side and defeat for the other. Instead, finding common ground and being willing to make concessions can lead to mutually beneficial outcomes. This requires a shift from a competitive mindset to one that values collaboration and collective well-being.

Moreover, adopting a solution-oriented approach can transform conflicts into opportunities for innovation and creativity. By focusing on potential solutions rather than dwelling on problems, we open the door to new possibilities. This proactive stance encourages constructive dialogue and reinforces the notion that conflicts, when handled with grace, can lead to positive change.

Additionally, it is important to reflect on conflicts after they have been resolved. Analyzing what worked and what didn't can provide valuable insights for future interactions. This reflection fosters a growth mindset, allowing individuals to learn from their experiences and refine their conflict-resolution skills over time.

Ultimately, navigating conflicts with grace requires a commitment to personal development and a desire to foster harmonious relationships. It involves cultivating patience, practicing self-awareness, and embracing vulnerability. By approaching conflicts with an open heart and a willingness to understand, we not only resolve disputes but also strengthen the bonds that connect us.

In the grand tapestry of life, conflicts are but threads woven into our experiences. By handling them with grace, we transform these threads from potential tangles into intricate patterns of understanding and growth. Thus, we fail without shame but with dignity and resilience, emerging wiser and more connected to those around us.

Learning from Relationship Failures

In the intricate tapestry of human interactions, relationships stand as both our greatest teachers and our most profound challenges. When a relationship falters, it often leaves us with a sense of inadequacy, a feeling of being unworthy of love or companionship. Yet, within these very failures lies the potential for profound personal growth and self-discovery.

Consider the notion that every relationship that ends is not merely a failure, but a steppingstone towards understanding ourselves better. Each failed relationship offers a mirror reflecting our insecurities, our unaddressed fears, and our unmet needs. It is in these reflections that we find the opportunity to learn and grow.

When a relationship fails, it is natural to feel a sense of loss and disappointment. However, within this emotional turmoil lies the chance to reevaluate our values and priorities. Were we truly aligned with our partner, or were we compromising essential aspects of ourselves for the sake of the relationship? The dissolution of a relationship can prompt us to examine these questions critically, leading to a deeper understanding of what we genuinely seek in a partner.

Moreover, relationship failures invite us to confront our communication patterns. How often do we find ourselves caught in cycles of misunderstanding, unable to express our true feelings or needs? By examining these patterns, we can identify areas where we may need to develop more effective communication skills. This self-awareness empowers us to engage in healthier, more fulfilling relationships in the future.

Another crucial lesson from failed relationships is the importance of boundaries. In the pursuit of love and connection, we may sometimes allow our boundaries to blur, leading to resentment and dissatisfaction. Recognizing this tendency can help us establish clearer boundaries in future relationships, ensuring that we maintain our sense of self while also nurturing a connection with another.

It is also essential to acknowledge the role of personal accountability in relationship failures. Blaming our partner for the breakup may provide temporary relief, but it ultimately hinders our growth. By taking responsibility for our actions and recognizing our contributions to the relationship's demise, we open ourselves to change and improvement. This accountability is not about self-blame but rather about empowering ourselves to make better choices moving forward.

Furthermore, failed relationships often highlight the necessity of self-love and self-compassion. When we experience rejection or betrayal, it is easy to internalize these experiences as evidence of our unworthiness. However, by cultivating a sense of self-love, we can begin to see ourselves as deserving of healthy, reciprocal relationships. This shift in perspective allows us to approach future relationships from a place of confidence and self-assurance.

Ultimately, learning from relationship failures requires a willingness to embrace vulnerability and an openness to change. It demands that we view each ending not as a defeat but as a valuable lesson that enriches our understanding of love and connection. By reframing our perspective on failure, we can transform these experiences into opportunities for growth, paving the way for more meaningful and fulfilling relationships in the future.

Building Stronger Bonds

A life without failure is a life without growth, and growth thrives in the fertile soil of relationships. Whether it be family, friendships, or professional connections, the bonds we forge with others are the scaffolding that supports us through our most challenging trials. Yet, the fear of failure often erects barriers that prevent us from forming these vital connections. To build stronger bonds, we must first dismantle the walls of shame that isolate us.

Shame whispers that vulnerability is weakness, that admitting our failures makes us less worthy of love and respect. But vulnerability is not a crack in our armor; it is the very essence of human connection. It is in our raw, unguarded moments that we truly touch the hearts of others. Sharing our struggles and setbacks is an invitation for empathy and understanding, a testament to our shared human experience. By revealing our imperfections, we give others permission to do the same.

The path to stronger bonds begins with active listening. To truly connect with someone, we must listen not just with our ears but with our hearts. This means being fully present in conversations, setting aside distractions, and engaging with genuine curiosity. When we listen actively, we honor the person speaking, and in doing so, we lay the foundation for mutual trust. Trust is the bedrock of any relationship, and without it, bonds cannot flourish.

Open communication further fortifies our connections. It involves more than just speaking honestly; it requires us to express our feelings and needs clearly and respectfully. When we communicate openly, we eliminate the guesswork that often leads to misunderstandings and resentment. It is a courageous act to articulate our desires and fears, but in doing so, we create a space where others feel safe to share their truths as well.

Empathy is the bridge that spans the divide between individuals. It allows us to step into another's shoes, to see the world through their eyes. Empathy is not about solving problems or offering advice; it is about being present, acknowledging someone else's reality, and validating their emotions. When we practice empathy, we cultivate compassion, which strengthens our bonds immeasurably.

Conflict is an inevitable part of any relationship, but it need not be destructive. When approached with a mindset of growth and understanding, conflict can be a catalyst for deeper connection. It is an opportunity to learn more about ourselves and others, to refine our communication skills, and to clarify our values and boundaries. By addressing conflicts with an open mind and a willingness to compromise, we reinforce the resilience of our bonds.

Building stronger bonds is not a one-time effort but a continuous process. It requires patience, persistence, and a commitment to nurturing the relationships that matter most to us. It involves celebrating successes together and supporting each other through failures. In doing so, we create a network of support that not only withstands the storms of failure but thrives because of them.

In the end, it is our connections with others that give our lives meaning. By rejecting shame and embracing vulnerability, by listening, communicating, empathizing, and resolving conflicts with grace, we build bonds that are unbreakable. These bonds become our most valuable assets, enriching our lives in ways that success alone never could.

Eight

OVERCOMING
PROFESSIONAL
SETBACKS

Career Failures and Growth

Failure is an inevitable part of any career path, a truth that many of us find difficult to accept. Society often views failure with disdain, associating it with incompetence or lack of effort. Yet, the reality is that failure can be a powerful catalyst for growth and innovation. It is through our missteps and setbacks that we learn the most about ourselves, our capabilities, and the world around us.

Consider the stories of successful individuals who have faced significant career failures. They did not let these setbacks define them, but rather used them as steppingstones to propel themselves to greater heights. Failure, when viewed through the lens of opportunity, becomes a teacher, offering lessons that success could never impart. It challenges us to reassess our strategies, refine our skills, and, most importantly, build resilience.

Resilience is the cornerstone of growth. It is the ability to rise after a fall, to persist in the face of adversity. The workplace is a dynamic environment where change is constant and challenges are abundant. Those who learn to navigate these challenges with courage and determination are the ones who ultimately thrive. Failure tests our resilience, pushing us to develop a mindset that is not only open to change but actively seeks it out.

Moreover, failure fosters creativity and innovation. When traditional approaches falter, we are forced to think outside the box, to explore uncharted territories. This process of exploration and experimentation can lead to groundbreaking ideas and solutions that might never have emerged in a risk-averse environment. Failure encourages us to question the status quo and to envision new possibilities, driving us to innovate and evolve.

Additionally, embracing failure cultivates humility and empathy. It reminds us that we are not infallible, that we have much to learn from others. This humility opens the door to collaboration and teamwork, as we become more willing to seek help and share knowledge. Empathy grows as we understand that everyone faces challenges and setbacks, fostering a supportive and inclusive work culture.

In the professional world, learning from failure is not just beneficial but essential. It is a critical component of personal and career development. Those who fear failure often miss out on opportunities for growth, as they avoid taking risks or trying new things. On the other hand, those who accept failure as a natural part of the learning process are more likely to take bold steps and achieve remarkable success.

Therefore, it is time to redefine our relationship with failure. Instead of viewing it as a source of shame, we should see it as an integral part of our journey towards success. By shifting our perspective, we can transform failure from a paralyzing fear into a powerful motivator. We must encourage a culture where failure is not stigmatized but celebrated as a valuable learning experience.

Ultimately, the key to overcoming career failures lies in our ability to adapt, learn, and grow. By embracing failure without shame, we unlock our potential to innovate, to lead, and to make meaningful contributions to our fields. Failure is not the end, but rather the beginning of a new chapter in our professional lives, filled with opportunities for growth and transformation.

Embracing Change in the Workplace

Change is the heartbeat of growth, the pulse that propels organizations forward, transforming potential into reality. Yet, the very essence of change is often met with apprehension, a natural human response to the unknown. In the workplace, this apprehension can manifest as resistance, a barrier to progress that can stifle innovation and impede success. However, when approached with the right mindset, change becomes more than a challenge; it becomes an opportunity for reinvention and improvement.

The first step in reshaping our perception of change is to understand its inevitability. In a world that evolves at an unprecedented pace, clinging to the status quo is akin to standing still in a fast-moving stream. Organizations that thrive are those that not only acknowledge change but actively seek it out, recognizing that adaptability is the cornerstone of resilience. By fostering a culture that values flexibility, companies can create an environment where change is not feared but welcomed as a catalyst for positive transformation.

A pivotal element in embracing change is communication. Transparent and consistent communication mitigates uncertainty and builds trust among team members. When leaders articulate the reasons behind a change and its anticipated benefits, employees are more likely to feel involved and valued in the process. This sense of inclusion can significantly reduce resistance, transforming skepticism into support. Moreover, effective communication provides a platform for employees to voice their concerns and contribute their insights, enriching the change process with diverse perspectives and innovative solutions.

Equally important is the role of leadership in navigating change. Leaders set the tone for how change is perceived and implemented within an organization. By demonstrating a commitment to change through their actions and decisions, leaders can inspire confidence and motivate their teams to embrace new directions. This involves not only championing change initiatives but also providing the necessary resources and training to equip employees for success in a transformed environment.

Emotional intelligence plays a crucial role in this dynamic. Leaders and team members alike must cultivate empathy and understanding, acknowledging that change can be a source of stress and anxiety. By fostering a supportive atmosphere where emotions are recognized and addressed, organizations can ease the transition and maintain morale. This emotional support is complemented by practical strategies, such as setting clear goals, celebrating milestones, and recognizing the contributions of individuals throughout the process.

Ultimately, embracing change in the workplace is about shifting the narrative from fear to opportunity. It requires a collective mindset that views change as an integral part of the journey towards excellence. By cultivating a culture that encourages experimentation, values feedback, and rewards adaptability, organizations can harness the power of change to drive innovation and achieve lasting success.

In this ever-evolving landscape, the ability to adapt is not just an asset; it is a necessity. As employees and leaders alike learn to navigate the complexities of change with confidence and creativity, they pave the way for a future where failure is not a source of shame but a steppingstones to greater achievements. By embracing the transformative potential of change, the workplace becomes a dynamic space of growth, learning, and limitless possibilities.

Rebuilding After Job Loss

Finding oneself suddenly without employment can feel like a storm has swept through one's life, leaving behind uncertainty and fear. Yet, it is precisely in these moments of upheaval that the seeds of opportunity begin to germinate. It is crucial to acknowledge that although the loss of a job is a significant event, it does not define one's worth or potential. Instead, it can serve as a catalyst for growth and transformation.

The first step in the rebuilding process is to confront the emotional turmoil head-on. Allow yourself to feel the full spectrum of emotions, from anger and sadness to relief and hope. Denying these feelings only prolongs the healing process. Acknowledge them, accept them, and let them guide you towards a deeper understanding of your inner self. This self-awareness is the foundation upon which you will build your new path.

Consider this an opportunity to reassess your career goals and aspirations. Often, we become so entrenched in our professional roles that we lose sight of our true passions and interests. Use this time to explore new avenues and potential career paths that align more closely with your personal values and long-term objectives. This period of reflection is not merely a pause but a powerful pivot towards a more fulfilling future.

Networking becomes an invaluable tool in this phase. Reach out to former colleagues, mentors, and industry contacts, not just to seek employment but to gather insights and advice. These connections can provide valuable support and may open doors to opportunities you hadn't previously considered. In today's interconnected world, networking is not just about finding a job but about building a community that can support and inspire you.

Skill enhancement is another critical component of rebuilding. The job market is ever-evolving, and staying competitive requires continuous learning. Identify skills that are in demand within your industry or explore new fields that pique your interest. Online courses, workshops, and certifications are readily available resources to help you upskill and stay relevant.

Moreover, embrace adaptability as a core strength. The ability to pivot and adjust to new circumstances is a highly valued trait in any professional landscape. By demonstrating resilience and flexibility, you not only enhance your employability but also cultivate a mindset that thrives in the face of change.

Financial stability is a concern for many who experience job loss. Crafting a realistic budget and exploring temporary employment options can ease the immediate pressure. Consider freelance opportunities or part-time roles that can provide income while you focus on your long-term career goals.

Above all, maintain a positive outlook. Surround yourself with supportive individuals who uplift and encourage you. Their belief in your potential can bolster your confidence and motivate you to persevere. Remember, the path to rebuilding is not linear; it is a series of steps, each bringing you closer to the life you envision.

Ultimately, the loss of a job can be a transformative experience. It is an invitation to redefine success on your own terms and to pursue a career that is not only financially rewarding but also personally fulfilling. Embrace this chapter with courage and conviction, knowing that your greatest achievements often arise from the ashes of adversity.

Finding New Paths

We often find ourselves trapped in the cycle of shame when facing failure, feeling as though there is no way forward. However, the key to breaking free from this cycle lies not in dwelling on what went wrong but in daring to chart new courses. By redefining our concept of success and failure, we can transform setbacks into opportunities for growth and innovation.

Consider the notion that failure is not the opposite of success but rather a steppingstones towards it. This perspective shift is crucial. Instead of viewing failure as an endpoint, see it as a valuable learning experience, a chance to gather insights that can guide you onto new paths you might not have considered before. This approach requires courage, as it challenges the conventional narratives that equate success solely with achievement and perfection.

When faced with failure, pause to reflect on the lessons it offers. What did you learn about your strengths, your weaknesses, and your resilience? Use these insights to inform your next moves. This process is not about clinging to past mistakes but using them as a springboard for innovation. The world is full of examples where initial failures paved the way for groundbreaking success—think of inventors, entrepreneurs, and artists who, undeterred by setbacks, discovered new avenues that led to their ultimate triumphs.

To find new paths, it is essential to cultivate a mindset that welcomes experimentation and risk-taking. This means being open to trying new things, even if they might lead to further failures. Every attempt, successful or not, enriches your understanding and expands your skill set. By fostering a culture of experimentation, you allow yourself the freedom to explore, to pivot when necessary, and to discover what truly resonates with your passions and capabilities.

Moreover, finding new paths often involves seeking out different perspectives and collaborating with others. Engage with people from diverse backgrounds and disciplines. Their insights can illuminate possibilities you might have overlooked. Collaboration can also provide support and encouragement, helping to mitigate the feelings of isolation that often accompany failure.

It's equally important to set realistic goals and celebrate small victories along the way. These incremental successes build confidence and momentum, propelling you forward on your new path. Recognize that progress is not always linear; setbacks may occur, but each step forward, no matter how small, is a testament to your resilience and determination.

Lastly, adopt a forward-thinking mindset. Focus on where you want to go, rather than where you have been. Let your vision of the future guide your actions and decisions. This doesn't mean ignoring the past, but rather integrating its lessons to forge a brighter path ahead.

In embracing failure as a natural part of our journey, we empower ourselves to pursue new directions with vigor and creativity. By redefining failure as an opportunity for growth, we not only dispel the shame associated with it but also unlock our potential to achieve remarkable success. This is the transformative power of finding new paths.

Nine

CULTIVATING PARENTING AND FAILURE

Teaching Resilience

R esilience is not merely a trait, it is a skill, and like any skill, it can be taught, honed, and perfected. In a world that often equates failure with inadequacy, fostering resilience becomes essential for personal growth and success. Resilience is the armor that shields us from the corrosive effects of shame and self-doubt, allowing us to confront setbacks with a fortified spirit.

The impact of resilience on an individual's ability to navigate life's challenges cannot be overstated. It transforms failure from a source of shame into a steppingstone for growth. When students, employees, or anyone in a learning environment are taught resilience, they learn to view failure as a temporary state rather than a definitive end. This perspective shift is crucial in dismantling the stigma associated with failure.

Imagine a classroom where mistakes are celebrated as opportunities for learning, where each error is a chance to refine one's approach rather than a reason to retreat. In such an environment, resilience flourishes. Students learn to analyze their missteps, understand the underlying causes, and strategize improvements—skills that extend far beyond the classroom and into every aspect of life.

The process of teaching resilience involves several key elements. First, it requires creating a safe space where individuals feel comfortable taking risks without the fear of judgment. Encouragement and support from educators, peers, and mentors play a pivotal role. By fostering a community that values effort and perseverance over perfection, we lay the foundation for resilience.

Moreover, teaching resilience necessitates the introduction of strategies for managing stress and adversity. Techniques such as mindfulness, positive self-talk, and reflection help individuals maintain a balanced perspective during challenging times. These tools equip them to cope with setbacks in a healthy manner, reducing the likelihood of succumbing to shame or self-doubt.

Role models also significantly influence the development of resilience. Witnessing others navigate failures with grace and determination provides a powerful example for learners. Stories of successful individuals who have faced and overcome significant obstacles can inspire and motivate, serving as a reminder that failure is not the end but a part of the journey towards achievement.

Additionally, resilience is bolstered by a growth mindset—a belief that abilities can be developed through dedication and hard work. This mindset encourages individuals to embrace challenges, persist in the face of setbacks, and see effort as a path to mastery. By instilling a growth mindset, we empower individuals to view failure as a natural and essential part of the learning process.

In teaching resilience, we do not merely equip individuals to withstand life's storms; we empower them to thrive despite them. Resilience transforms potential pitfalls into platforms for growth, fostering a sense of agency and confidence. It enables individuals to approach each new challenge with optimism and determination, knowing that failure is not a reflection of their worth but an opportunity for improvement.

Ultimately, teaching resilience is about cultivating an unwavering belief in one's ability to overcome obstacles. It is about nurturing a mindset that sees failure not as a source of shame, but as a catalyst for strength and self-discovery. By embedding resilience into the fabric of education and personal development, we create a society where failure is no longer feared but embraced as a vital component of success.

Encouraging Growth Through Mistakes

Imagine a world where every misstep is not just tolerated but celebrated as a steppinstone to greatness. This is not a utopian vision; it's a transformative mindset that can propel individuals to heights they never dreamed possible. The truth is, mistakes are not the enemy of progress; they are its catalyst. Every failure contains within it the seeds of innovation, the spark of creativity, and the blueprint for personal growth.

Too often, society conditions us to view mistakes as weaknesses, blemishes on the otherwise pristine canvas of our lives. This perspective is not only limiting but also detrimental. It stifles creativity, curtails ambition, and breeds a culture of fear. Instead, imagine flipping the narrative. Viewing mistakes as valuable experiences can be a game-changer, fostering an environment ripe for learning and development.

When we make a mistake, it forces us to pause and reflect. This reflection is where the magic happens. It allows us to dissect what went wrong, understand why it happened, and devise strategies to avoid similar pitfalls in the future. This process of analysis and adaptation is the essence of learning. It transforms setbacks into powerful lessons that can inform future decisions, making us more resilient and better equipped to handle challenges.

Moreover, mistakes encourage humility and vulnerability, essential traits for personal growth. Acknowledging our mistakes requires us to confront our limitations and biases, opening the door to self-improvement. It teaches us to accept constructive criticism and seek guidance from others, fostering an environment of collaboration and mutual support. This openness to learning from our errors enriches our perspective and enhances our capacity to innovate.

Consider the stories of great inventors, scientists, and leaders throughout history. Many of them attribute their success to the lessons learned from their failures. Thomas Edison's countless unsuccessful attempts at inventing the lightbulb were not failures in his eyes but rather essential steps toward his ultimate success. Each failure was a lesson, a chance to iterate and refine his approach. This mindset not only led to his monumental achievements but also laid the foundation for countless innovations that followed.

In the realm of personal development, mistakes serve as a mirror reflecting areas for improvement. They illuminate our weaknesses, allowing us to focus our efforts on overcoming them. This process of continuous self-assessment and growth is vital for achieving our full potential. By embracing our mistakes, we cultivate a growth mindset, one that sees challenges as opportunities rather than threats.

Encouraging growth through mistakes is not about glorifying failure but about recognizing its value in the journey toward success. It's about fostering a culture that sees mistakes as an integral part of the learning process, not as an endpoint. This shift in perspective can unleash untapped potential, drive innovation, and pave the way for extraordinary achievements.

Let us redefine failure, not as a sign of defeat but as a badge of honor, a testament to our courage to try, to risk, and to learn. By doing so, we can transform our approach to mistakes, turning them into powerful catalysts for growth and success. In this new paradigm, failure is not something to be ashamed of but a vital part of the path to greatness.

Balancing Protection and Freedom

In the intricate dance of life, we often find ourselves caught between two opposing forces: the need for protection and the desire for freedom. This delicate balance is pivotal in navigating the tumultuous seas of failure and success without succumbing to shame. It's about understanding how to shield ourselves from undue harm while simultaneously granting ourselves the liberty to explore, to err, and to grow.

Protection provides a cocoon, a safe haven where we can retreat when the world feels too harsh. It is the safety net that catches us when we fall, offering solace and security. Yet, if held too tightly, protection can become a cage, stifling our growth and muting the vibrant colors of our potential. It whispers caution, urging us to avoid risk and steer clear of failure, which, ironically, are the very elements essential for true learning and development.

Freedom, on the other hand, is the open sky, inviting us to spread our wings and soar. It is the exhilarating rush of possibility and the thrill of the unknown. Freedom encourages us to take risks, to step outside our comfort zones, and to embrace the beauty of imperfection. However, without the tempering influence of protection, this freedom can lead to reckless decisions and unnecessary harm.

The key lies in finding equilibrium between these two forces. This balance is not a static state but a dynamic process, constantly shifting with our circumstances and growth. It requires self-awareness and the courage to reassess our boundaries and beliefs. By cultivating this balance, we can learn to fail without shame, understanding that failure is not a reflection of our worth but a steppingstone towards greater wisdom and resilience.

To achieve this balance, we must first redefine our perception of failure. Instead of seeing it as a definitive end, we should view it as a valuable teacher. Failure's lessons are often more profound than those of success, offering insights into our strengths and weaknesses, and helping us refine our paths. By embracing failure as a natural and necessary part of life, we can mitigate the shame that often accompanies it, allowing ourselves to take risks and pursue our passions without fear of judgment.

Moreover, we must learn to establish boundaries that protect us from undue harm while allowing for growth and exploration. This involves setting realistic expectations, seeking support when needed, and being mindful of our emotional and physical well-being. Protection should not be about avoiding failure altogether but about creating a supportive environment where we can learn from our mistakes and continue to evolve.

In this delicate balance, communication plays a crucial role. Sharing our experiences and struggles with others can provide perspective and encouragement. It reminds us that we are not alone in our journey and that others have faced similar challenges and emerged stronger.

Ultimately, balancing protection and freedom is an ongoing journey of self-discovery and growth. It requires us to be both vulnerable and resilient, to guard our hearts while keeping them open to the world. By mastering this balance, we empower ourselves to navigate the complexities of life with grace and courage, transforming failure from a source of shame into a catalyst for change and self-improvement.

Nurturing a Growth Mindset

Imagine a world where every setback is seen not as a reflection of one's inherent limitations but as an opportunity for growth. This perspective transforms not only how we perceive failure but also how we engage with challenges. It is the foundational difference between a fixed mindset and a growth mindset—the belief that abilities and intelligence can be developed with effort, learning, and dedication.

The journey towards nurturing a growth mindset begins with recognizing the power of "yet." When faced with a challenge or a failure, the addition of this small word shifts the narrative from "I can't do it" to "I can't do it yet." This simple linguistic shift opens the door to possibility and potential, allowing individuals to view their abilities as elastic rather than static. It encourages a focus on learning and improvement, fostering resilience and a willingness to take risks.

Consider the transformative power of re-framing mistakes as steppingstones rather than stumbling blocks. When we view errors as integral components of the learning process, we cultivate an environment where experimentation is encouraged and innovation thrives. This perspective is vital in a world that often equates success with perfection. By embracing imperfections and setbacks as opportunities for growth, individuals are more likely to persevere and ultimately achieve greater success.

The cultivation of a growth mindset also requires a shift in how we interpret feedback. Instead of perceiving criticism as a personal attack, it should be seen as valuable information that guides improvement. Constructive feedback becomes a tool for development, not a confirmation of inadequacy. This approach empowers individuals to seek out and learn from feedback, fostering a culture of continuous improvement.

Moreover, nurturing a growth mindset involves celebrating effort and perseverance rather than just innate talent or intelligence. By recognizing and rewarding hard work and determination, we can inspire individuals to push beyond their current capabilities. This not only boosts self-esteem but also reinforces the idea that effort leads to achievement.

The role of self-talk cannot be underestimated in this process. The internal dialogue we maintain shapes our beliefs about our abilities. Encouraging positive self-talk that emphasizes growth and potential can significantly influence one's mindset. By replacing negative, limiting thoughts with affirmations of capability and resilience, we lay the groundwork for a mindset that thrives on challenges.

It's important to remember that nurturing a growth mindset is not a one-time effort but a continuous journey. It requires ongoing reflection, practice, and a willingness to embrace discomfort. As we consistently challenge ourselves to step out of our comfort zones and approach setbacks with curiosity instead of fear, we gradually embed the principles of a growth mindset into our daily lives.

In a world that often prioritizes immediate success and measurable outcomes, the value of a growth mindset cannot be overstated. It empowers individuals to redefine failure as a natural part of the learning process and to pursue their goals with renewed vigor and hope. By nurturing this mindset, we unlock the potential to fail without shame, transforming each setback into a steppingstone towards greater achievements. In doing so, we create a culture where growth is celebrated, and the pursuit of excellence is fueled by perseverance and an unwavering belief in the power of potential.

Ten

SPIRITUAL PERSPECTIVES ON FAILURE

Inner Peace A Midst Chaos

In the midst of our most tumultuous times, the concept of failure often looms large, casting shadows of doubt and anxiety over our every endeavor. Yet, it is in these very moments that the opportunity for true growth and understanding presents itself. The chaos that surrounds us is not a barrier, but rather a catalyst for finding the inner peace that resides within us all.

To navigate the stormy seas of failure without the burden of shame, one must first recognize that failure is not the antithesis of success, but a vital component of the journey toward it. Embracing this perspective allows us to strip away the layers of self-doubt and societal pressure that often accompany perceived shortcomings. Instead, we can culti-vate a mindset that sees each misstep not as a definitive end, but as a valuable lesson in the ongoing process of personal development.

The world around us is filled with noise—expectations, judgments, and the constant hum of comparison. Yet, a midst this clamor, there exists a profound silence, a space within where peace can be found. It is here, in the quiet recesses of our minds, that we can begin to dismantle the notion that failure is synonymous with inadequacy. By re-framing our understanding of what it means to fail, we open ourselves to the possibility of growth that is unencumbered by shame.

Consider the act of failure as a mirror reflecting the areas where we have room to grow. Each setback, each stumble, is an invitation to pause, reflect, and re-calibrate. This process of introspection, though often uncomfortable, is where the seeds of resilience are sown. By confronting our fears and acknowledging our vulnerabilities, we empower ourselves to move forward with renewed purpose and clarity.

The key to achieving inner peace a midst chaos lies in our ability to remain anchored in the present moment. Mindfulness practices, such as meditation and focused breathing, serve as powerful tools in cultivating a sense of calm and balance. These practices allow us to detach from the external chaos and reconnect with the stillness within. In doing so, we become better equipped to face challenges with a clear mind and an open heart.

It is also essential to foster a sense of self-compassion. We are often our harshest critics, quick to judge ourselves for perceived failures. However, by treating ourselves with kindness and understanding, we create a nurturing environment in which we can flourish. Self-compassion is the antidote to shame, allowing us to acknowledge our mistakes without internalizing them as a reflection of our worth.

As we navigate the complexities of life, let us remember that inner peace is not a destination, but a state of being. It is the quiet confidence that arises from knowing that failure is not a reflection of our inadequacy, but a testament to our courage to try, to learn, and to grow. In this way, we can move through the chaos with grace and resilience, unburdened by shame and emboldened by the knowledge that we are more than the sum of our failures. In finding peace within, we discover the strength to face whatever challenges lie ahead.

Finding Meaning in Adversity

In an era where perfection is paraded as the pinnacle of success, admitting failure often feels like a public confession of inadequacy. Yet, it is in the throes of adversity that we discover the profound depths of human resilience and the true essence of personal growth. Adversity, often perceived as an unwelcome guest, can be a powerful catalyst for transformation when we allow ourselves to see beyond its initial sting.

Imagine for a moment standing at the edge of a vast, unknown wilderness. The path behind is familiar, safe, but the unknown before you beckons with the promise of untapped potential. This is the landscape of adversity. When faced with challenges, our instinct is to retreat, to shield ourselves from discomfort. However, it is precisely in these moments of vulnerability that we can forge a deeper understanding of ourselves and our capabilities.

Adversity strips away the superficial layers of our identity, leaving us with the raw materials of our character. It forces us to confront our limitations, fears, and insecurities, demanding honesty and introspection. This process, though uncomfortable, is crucial for authentic self-discovery. As we grapple with our shortcomings, we develop a clearer vision of our strengths, values, and desires.

Moreover, adversity teaches us the invaluable lesson of perseverance. It is easy to remain steadfast when the path is smooth and the horizon clear, but true resilience is born in the crucible of hardship. Every challenge surmounted, every setback endured, adds another layer of fortitude to our spirit. We learn to push forward, not in spite of adversity, but because of it, recognizing that each obstacle is an opportunity to hone our resolve.

Embracing adversity also fosters empathy and compassion. When we experience struggles, we gain a deeper understanding of the challenges faced by others. This shared experience of hardship creates a profound sense of connection and community. We learn to extend grace and support to those around us, recognizing that we are all navigating our own battles. In this way, adversity not only strengthens our individual resolve but also enriches our collective humanity.

Finding meaning in adversity requires a shift in perspective. Instead of viewing challenges as insurmountable barriers, we can choose to see them as stepping stones towards greater wisdom and strength. This mindset does not diminish the pain of failure but rather reframes it as a vital component of personal evolution. By transforming our understanding of adversity, we empower ourselves to face the future with courage and confidence.

In the labyrinth of life, failure is not the end but a necessary detour on the path to fulfillment. By embracing the lessons embedded in our struggles, we unlock the potential for profound personal growth. The journey through adversity is not a solitary endeavor but a shared human experience that binds us together in our pursuit of a meaningful existence. Let us not shy away from the trials that life presents but rather embrace them as opportunities to discover the limitless depths of our potential.

Failure and The Divine

In the tapestry of life, where threads of success and failure are woven together, the concept of divine intervention often emerges as a beacon of hope and understanding. The notion that a higher power has a hand in our triumphs and setbacks can offer solace, perspective, and a deeper connection to life's purpose.

Failure, often perceived as a dark cloud, can obscure the light of potential. Yet, when viewed through the lens of the divine, it transforms into a pivotal moment of growth and learning. It is a sacred pause, an invitation to introspection, urging us to seek the lessons embedded within the struggle. The divine narrative suggests that failure is not an endpoint but a stepping stone to greater enlightenment.

Consider the stories of those who have faced monumental failures yet emerged stronger, wiser, and more attuned to their life's calling. These individuals often attribute their resilience and newfound direction to a belief in a greater plan. This belief acts as a guiding compass, steering them through the tumultuous seas of doubt and despair, toward a horizon filled with promise. It is within this framework of understanding that failure sheds its negative connotations and reveals its true essence—a divinely orchestrated opportunity for transformation.

The divine perspective encourages us to view failure not as a reflection of our inadequacy but as a testament to our potential. It teaches us that the divine does not measure success in the same way society does. Instead, it values perseverance, humility, and the courage to rise after each fall. By embracing this viewpoint, we begin to see failure as a necessary and valuable component of our spiritual and personal development.

Moreover, the divine narrative offers a profound sense of companionship. In moments of failure, when isolation and self-doubt threaten to overtake us, the belief in a supportive, divine presence can provide comfort and reassurance. It whispers the timeless truth that we are never alone on our path. Our struggles are not ours to bear alone but are shared with a compassionate force that urges us forward, even when the way seems unclear.

This divine relationship with failure also cultivates gratitude. It encourages us to appreciate the growth that emerges from adversity and to recognize the hidden blessings within our challenges. When we perceive failure as a gift, wrapped in divine intention, our hearts open to a deeper appreciation for the journey of life itself.

As we navigate the complexities of success and failure, embracing the divine perspective allows us to rewrite our narratives. We move beyond the fear of failure and towards a place of acceptance and peace. By acknowledging the divine's role in our lives, we cultivate resilience and an unwavering belief in our ability to rise above any obstacle.

In this light, failure becomes a sacred teacher, guiding us to a more profound understanding of ourselves and our place in the world. It is a divine dance, where we learn to trust in the unfolding of our lives, knowing that each step, whether forward or backward, is purposeful and meaningful. Through this lens, we find the courage to fail without shame, embracing the divine wisdom that accompanies every fall.

Shifting Perspectives from Shame

Imagine standing at the edge of a vast, daunting cliff, peering down into the abyss below. This cliff represents the overwhelming sense of shame that many feel when confronted with failure. The fear of falling into that abyss often paralyzes us, preventing us from taking the steps necessary to move forward. But what if we could shift our perspective and see not a terrifying drop but an opportunity to soar?

Shame, by its very nature, is a deeply ingrained emotion that thrives in the shadows of secrecy and silence. It whispers lies about our worth, convincing us that our mistakes define us. However, this narrative can be rewritten. The first step in dismantling shame's hold is to question its validity. Ask yourself: Is this feeling of shame truly justified, or is it a learned reaction from societal pressures and internalized expectations?

The power of perspective cannot be underestimated. By reframing our thoughts around failure, we create space for growth and resilience. Consider failure not as a reflection of inadequacy but as an integral part of the learning process. Each misstep is a stepping stone toward mastery, a chance to refine our skills and deepen our understanding. When we view failure through this lens, shame loses its sting and becomes a catalyst for progress.

Moreover, embracing vulnerability is crucial. It may seem counterintuitive, but acknowledging our imperfections and sharing them with others fosters a sense of connection and empathy. This openness can dismantle the isolating walls of shame, allowing us to find strength in our shared humanity. By cultivating a supportive environment, we empower ourselves and others to face challenges with courage and compassion.

Another transformative approach is to focus on self-compassion. Treat yourself with the same kindness and understanding you would offer a friend. Recognize that everyone falters and that these experiences do not diminish your value. By nurturing a gentle inner dialogue, you can build resilience against shame's corrosive effects. This shift in mindset encourages a healthier relationship with failure, where mistakes are seen as opportunities for self-improvement rather than sources of self-condemnation.

In addition, challenge the narratives that society imposes. The cultural obsession with perfection and success is an unrealistic standard that often fuels shame. Instead, celebrate authenticity and diversity in experiences. By redefining success on your terms, you liberate yourself from the constraints of societal judgment.

Ultimately, shifting perspectives requires conscious effort and practice. It demands that we confront the discomfort of shame and transform it into a tool for empowerment. As you navigate the complexities of failure, remember that you hold the power to redefine your narrative. By choosing to see beyond the shame, you unlock the potential for growth and self-discovery.

In this journey of transformation, you are not alone. Countless others have walked this path, and their stories of resilience and triumph serve as beacons of hope. Together, we can dismantle the barriers of shame, embracing a future where failure is not feared but embraced as a stepping stone toward a more authentic and fulfilling life. Let this new perspective guide you, as you soar beyond the cliff's edge, unburdened by the weight of shame.

The Journey Toward Enlightenment

In a world where the fear of failure often casts a long shadow over our ambitions, the path to enlightenment stands as a beacon of hope and transformation. This path is not merely a destination but a profound shift in perspective—a realization that failure is not a final verdict but an essential stepping stone to growth and self-discovery.

Enlightenment in the context of failure begins with a fundamental reevaluation of our relationship with setbacks. Society has conditioned us to view failure as something to be avoided at all costs. Yet, it is through failure that the most profound lessons are learned. It is the crucible in which resilience is forged, creativity is ignited, and character is built. By reframing failure as an opportunity for growth, we unlock the potential to rise above our limitations.

The path to enlightenment also demands a deep understanding of the self. It requires introspection and the courage to confront our fears, doubts, and insecurities. This self-awareness is crucial, for it is only by acknowledging our vulnerabilities that we can begin to transcend them. Through this process, we cultivate a mindset that is unafraid to take risks and embrace uncertainty.

Moreover, the pursuit of enlightenment involves cultivating a sense of purpose that transcends individual aspirations. When we align our goals with a higher purpose, the fear of failure diminishes. Our efforts become part of a larger narrative, one that is not solely defined by personal success but by the impact we have on the world around us. This shift from self-centered ambition to a more altruistic perspective imbues our actions with meaning and resilience.

Community and connection play a vital role in this transformative journey. Surrounding ourselves with individuals who share our values and encourage our growth creates a supportive environment where failure is not met with judgment but with understanding and encouragement. It is within these communities that we find the strength to persevere, to learn from our mistakes, and to continue striving toward our goals.

The path to enlightenment is also marked by the cultivation of gratitude and mindfulness. By appreciating the present moment and recognizing the abundance in our lives, we shift our focus from what we lack to what we have. This gratitude nurtures a sense of contentment and peace, allowing us to face challenges with equanimity and grace.

Ultimately, the path to enlightenment is a journey of transformation. It is an ongoing process that requires patience, dedication, and an unwavering belief in the power of change. As we navigate this path, we discover that failure is not a barrier but a catalyst for growth. We learn to view setbacks not as defeats but as opportunities to evolve and become the best versions of ourselves.

In this journey, we find that enlightenment is not a distant destination but a state of being—a harmonious blend of self-awareness, purpose, and connection that empowers us to live authentically and fearlessly. As we embrace this path, we transcend the shame of failure and emerge stronger, wiser, and more compassionate, ready to contribute meaningfully to the world around us.

Eleven

BUILDING RESILIENCE THROUGH FAILURE

Developing Grit

I n the grand tapestry of personal growth, one thread stands out as the linchpin of success: grit. It is the steadfast resolve to persevere through challenges, an unwavering commitment that transforms failures into stepping stones. In a society that often stigmatizes failure, the cultivation of grit becomes not just a desirable trait, but a necessary one. It is the armor that shields us from the paralyzing fear of failure and propels us towards our goals with relentless determination.

In today's fast-paced world, where the pressure to succeed is immense, many individuals find themselves shackled by the fear of falling short. This fear, however, is not an insurmountable barrier. It is a call to cultivate grit, to nurture an inner tenacity that refuses to be defined by setbacks. Grit is not an innate quality; it is a skill that can be developed through practice and perseverance.

The first step in developing grit is to redefine failure. Viewing setbacks as opportunities for growth rather than as personal shortcomings can radically shift one's perspective. This mindset fosters resilience, encouraging individuals to learn from their mistakes and continue striving towards their objectives. By reframing failure as a natural and essential part of the learning process, we can shed the cloak of shame that often accompanies it.

Moreover, setting realistic and challenging goals is crucial in the grit-building process. These goals act as beacons, guiding us through the tumultuous seas of adversity. They provide a sense of direction and purpose, motivating us to push through obstacles with unwavering resolve. By breaking down larger ambitions into smaller, manageable tasks, we can maintain focus and avoid feeling overwhelmed.

Equally important is the cultivation of a growth mindset. This mindset, characterized by the belief that abilities and intelligence can be developed through dedication and hard work, is the fertile ground upon which grit flourishes. Embracing challenges, persisting in the face of setbacks, and seeing effort as a path to mastery are hallmarks of this mindset. It emboldens individuals to take calculated risks, knowing that failure is not the end, but a stepping stone towards success.

Support systems also play a vital role in nurturing grit. Surrounding oneself with individuals who encourage perseverance and resilience can provide the necessary reinforcement during difficult times. Mentors, peers, and family can offer guidance, share experiences, and provide the emotional support needed to stay the course.

Finally, self-compassion is an often-overlooked component in the development of grit. Being kind to oneself in times of failure fosters a healthier relationship with setbacks, allowing individuals to bounce back with renewed vigor. It is a reminder that failure is not a reflection of one's worth but a transient experience on the path to achievement.

Grit is not merely about enduring hardships; it is about transforming them into catalysts for growth. It is the relentless pursuit of one's passions, the refusal to be deterred by obstacles, and the courage to forge ahead despite the fear of failure. By developing grit, we equip ourselves with the tools to fail without shame, embracing each setback as a powerful teacher on the journey to success.

The Vital Role of Perseverance

Imagine standing at the crossroads of opportunity and failure. It's a place where dreams either take flight or wither away under the weight of setbacks. This juncture is familiar to all who dare to strive for greatness, and it's here that perseverance becomes not just a virtue, but a necessity.

Perseverance is more than a mere act of stubbornness; it is the lifeline that pulls individuals through the murky waters of self-doubt and adversity. It is the unwavering commitment to press on, even when the path forward is obscured by obstacles and fraught with uncertainty. History is replete with tales of those who, against all odds, refused to succumb to failure. Each of these stories serves as a testament to the power of persistence.

Consider the process of learning to ride a bicycle. The first attempts are often marked by falls and scrapes, yet it is the determination to get back up and try again that ultimately leads to success. In the same vein, life's challenges demand a similar resilience. It's not the avoidance of failure that defines success, but the courage to rise after each fall.

In the professional arena, perseverance can be the distinguishing factor between those who merely dream and those who achieve. The corporate world is a battlefield where setbacks are inevitable. Projects may falter, innovations may not always yield the desired results, and market conditions may shift unexpectedly. Yet, those who persist, who learn from each misstep and adapt their strategies, are the ones who carve out lasting legacies.

Perseverance also plays a crucial role in personal growth. It fosters the development of resilience and grit, qualities that are indispensable in navigating the complexities of life. By persistently pursuing goals, individuals cultivate a mindset that views challenges as opportunities for growth rather than insurmountable barriers. This shift in perspective can transform failures into stepping stones toward future success.

Moreover, perseverance teaches the invaluable lesson of patience. In a world that often glorifies instant gratification, the ability to remain steadfast in the pursuit of long-term goals is increasingly rare yet profoundly rewarding. It is through patience that one learns to appreciate the journey, savoring each small victory along the way.

Critically, perseverance is not a solitary endeavor. It thrives in environments that support and encourage resilience. Surrounding oneself with like-minded individuals who understand the importance of persistence can provide the motivation needed to keep moving forward. Whether through mentors, supportive peers, or inspiring role models, the collective energy of a community can propel individuals towards their goals.

In a society that often stigmatizes failure, perseverance stands as a beacon of hope. It challenges the notion that failure is something to be ashamed of, instead positioning it as an essential component of the learning process. By reframing failure as a temporary setback rather than a definitive end, perseverance empowers individuals to continue striving for their dreams.

Ultimately, perseverance is the bridge that connects aspiration with achievement. It is the force that transforms dreams into reality, the catalyst that turns potential into accomplishment. In the face of adversity, it is perseverance that whispers, "Keep going," ensuring that failure is not an end, but merely a stepping stone on the path to success.

Learning to Adapt

In the ever-evolving landscape of personal and professional life, the ability to adapt is not merely an asset but a necessity. As the world spins on its axis, we are constantly confronted with new challenges and unexpected turns. The instinctual response may be to resist, to cling to the familiar, but it is in these moments of discomfort that the seeds of growth are planted.

Adaptation, at its core, is about transformation. It is the art of recognizing the shifts around us and adjusting our sails accordingly. The narrative of failure is often tainted with the stigma of shame, a perception that can paralyze and impede progress. Yet, if we redefine failure not as a terminal endpoint but as a stepping stone to greater understanding, we unlock the potential for profound personal evolution.

Consider the story of the phoenix, a mythical creature that rises from its ashes, reborn and renewed. This imagery serves as a powerful metaphor for our own capacity for renewal. When faced with setbacks, the path forward is not to wallow in the ashes of what was but to use them as fertile ground for what can be. To adapt is to take stock of the lessons failure provides and to forge a new path with that knowledge.

The process of adaptation requires a mindset shift, a conscious decision to view every obstacle as an opportunity for learning. It demands the courage to step outside of comfort zones and the humility to acknowledge that we do not have all the answers. This mindset is the cornerstone of resilience, a quality that enables us to withstand the storms of life and emerge stronger on the other side.

Moreover, adaptability is intricately linked with creativity. In the face of adversity, the ability to think outside the box and devise innovative solutions is paramount. History is replete with examples of individuals who, when confronted with seemingly insurmountable challenges, harnessed their creative faculties to not only survive but thrive. Whether it is a business pivoting in response to market changes or an individual reassessing their personal goals, creative problem-solving is a vital component of successful adaptation.

To cultivate adaptability, one must also embrace the notion of continuous learning. The landscape of knowledge is ever-expanding, and those who remain curious and open-minded are better equipped to navigate change. This involves seeking out new experiences, engaging with diverse perspectives, and remaining receptive to feedback. By fostering a culture of learning, we enable ourselves to stay ahead of the curve and anticipate change rather than merely react to it.

In the pursuit of adapting without shame, it is crucial to foster a supportive environment. Surrounding oneself with a network of mentors, peers, and allies who encourage growth and provide constructive feedback is invaluable. Such a community acts as a buffer against the isolating nature of failure and reinforces the notion that adaptation is a shared journey.

Ultimately, learning to adapt is about embracing the fluidity of life. It is about acknowledging that change is the only constant and that our ability to navigate it with grace and resilience defines our success. In shedding the cloak of shame associated with failure and viewing it as an integral part of the growth process, we unlock our true potential and chart a course towards a future rich with possibilities.

Strategies for Bouncing Back

Picture this: you're standing on the edge of a cliff, staring down into the abyss of your failures, feeling the weight of shame like a boulder strapped to your back. It's a place we all find ourselves at some point, but it's what we do next that defines us. The key to transforming this moment of despair into a launchpad for success lies in adopting effective strategies to bounce back.

First, let's focus on reframing our mindset. Failure is not a reflection of your worth, but rather a stepping stone towards growth. By altering your perception of failure, you can begin to see it as a valuable teacher. This mindset shift empowers you to extract lessons from every setback. Analyze what went wrong, identify the factors within your control, and brainstorm actionable steps to prevent similar outcomes in the future. This cognitive restructuring is the foundation upon which resilience is built.

Next, leverage the power of self-compassion. It's easy to spiral into self-criticism, but harsh judgment only deepens the wounds of failure. Instead, treat yourself with the kindness you would offer a close friend. Acknowledge your efforts, recognize your courage to take risks, and remind yourself that imperfection is part of the human experience. By fostering a compassionate inner dialogue, you create a nurturing environment for recovery and renewal.

Now, consider the importance of setting realistic goals. In the aftermath of failure, it can be tempting to swing to the opposite extreme, setting overly ambitious objectives in an attempt to prove your worth. However, this often leads to burnout and further disappointment. Instead, break your goals into manageable milestones. Celebrate small victories along the way, as these incremental achievements build momentum and confidence, propelling you towards your ultimate aspirations.

Furthermore, cultivate a support network. Surround yourself with individuals who uplift and inspire you. Engaging with a community that shares similar experiences not only provides comfort but also offers diverse perspectives and solutions. Whether through mentorship, peer support groups, or professional networks, these connections can be a source of encouragement and accountability.

Another pivotal strategy is to embrace adaptability. The path to success is rarely linear, and the ability to pivot in response to obstacles is a crucial skill. Cultivate flexibility in your approach, remaining open to new ideas and methods. This adaptability not only enhances your problem-solving capabilities but also fosters an innovative mindset that thrives on change.

Finally, focus on self-care as an integral component of your recovery journey. Physical, emotional, and mental well-being are interconnected, and neglecting any aspect can hinder your ability to bounce back. Prioritize activities that replenish your energy, whether it's exercise, meditation, or creative pursuits. By maintaining a balanced lifestyle, you equip yourself with the resilience needed to face future challenges with vigor.

In the end, the ability to bounce back from failure is not about avoiding mistakes but about embracing them as catalysts for growth. It's about crafting a narrative where failure is not a dead end but a detour leading to richer, more fulfilling paths. By integrating these strategies into your life, you not only diminish the power of shame but also unlock your true potential, transforming setbacks into stepping stones towards success.

Developing Unshakable Confidence

Imagine standing on the precipice of a vast canyon, the wind tugging at your clothes, and the world sprawling infinitely before you. This is the moment where fear meets potential, where doubt transforms into certainty. Confidence is not a trait bestowed upon the fortunate few; it is a skill honed by those willing to step into the unknown, to fail, and to rise again.

Let us dispel the myth that confidence is an inherent quality, a gift wrapped at birth. Instead, view it as a muscle that grows stronger with each challenge faced, each fear conquered. The most assured individuals have not led lives free of failure; they have simply learned to view setbacks as stepping stones rather than stumbling blocks.

The first step towards developing unshakable confidence is understanding that failure is not a reflection of your worth, but a necessary component of growth. When you can stand in the aftermath of defeat and recognize its lessons, you begin to build a foundation of resilience. This resilience is the bedrock of confidence, a deep-seated belief that no matter the outcome, you have the strength to persevere.

Cultivating confidence requires a commitment to self-awareness. Take stock of your achievements, no matter how small. Each success is a testament to your capabilities, a reminder of what you can achieve when you dare to try. Celebrate these victories, for they are the building blocks of a confident mindset.

Next, consider the power of visualization. The mind is a powerful tool, capable of shaping reality through the imagery it creates. Picture yourself succeeding, imagine the feelings of accomplishment and pride that accompany your triumphs. This mental rehearsal prepares your mind for success, reinforcing the belief that you are capable of achieving your goals.

Surround yourself with positivity. The company you keep influences your perception of yourself. Engage with those who uplift and inspire you, who see your potential even when you cannot. Their belief in you will fuel your own, creating a supportive environment where confidence can flourish.

Moreover, embrace the power of preparation. Confidence often falters in the face of uncertainty. By arming yourself with knowledge and skills, you reduce the unknowns, converting fear into familiarity. Whether it is a presentation, a competition, or a conversation, preparation provides a safety net, allowing you to face challenges with assurance.

Finally, consider the impact of body language. Stand tall, make eye contact, and engage with the world around you. These physical manifestations of confidence can influence your internal state, creating a feedback loop that reinforces your belief in yourself.

The journey to unshakable confidence is not a straight path but a series of deliberate choices. It is the decision to rise after every fall, to learn from every misstep, and to believe in your inherent worth and capability. As you cultivate this confidence, remember that it is not the absence of fear but the mastery over it. With each step forward, you are not just building confidence; you are building a life unburdened by the fear of failure.

Twelve

REDEFINING SUCCESS ON YOUR TERMS

Personal Values and Goals

In the fast-paced world we live in, a relentless chase for perfection often overshadows the very essence of our personal values and goals. We are conditioned to equate success with societal norms, neglecting the unique paths that resonate with our true selves. But what if failing is not a source of shame but a pivotal step towards understanding who we are and what truly matters to us?

To navigate through life without the burden of shame, we must first unearth our core values—the principles that define us, that ignite our passions and guide our decisions. These values are the compass that directs us towards goals that are not only achievable but also meaningful. By aligning our aspirations with our values, we create a roadmap that is uniquely ours, free from the constraints of external validation.

Imagine a life where each goal you set is a reflection of your deepest convictions, where every step you take is fueled by genuine intention rather than obligation. In this paradigm, failure is not a reflection of inadequacy but a testament to your courage to pursue what truly matters. This perspective transforms failure into a powerful tool for growth, a catalyst for self-discovery.

When we set goals that are aligned with our values, we create a framework that supports resilience. This alignment fosters a sense of purpose and fulfillment, making the journey itself rewarding, regardless of the outcome. It liberates us from the fear of judgment and the pressure to conform, allowing us to embrace our authentic selves.

Consider the power of setting goals that are intrinsically motivated. These are goals that resonate with your inner desires, not imposed by societal expectations or fleeting trends. They are the goals that inspire creativity, enhance well-being, and cultivate a sense of satisfaction that transcends material achievements. When we pursue such goals, we become architects of our destiny, crafting a life that is rich and purposeful.

However, identifying personal values and setting aligned goals requires introspection and honesty. It demands that we peel away the layers of pretense and confront our true selves. This process may reveal uncomfortable truths, but it is through this vulnerability that we find strength. By acknowledging our imperfections, we open the door to growth and transformation.

The journey towards aligning personal values with goals is not without challenges. It requires patience, persistence, and a willingness to embrace uncertainty. Yet, it is in this very pursuit that we find liberation from the shackles of shame. We learn to celebrate our failures as milestones, each one bringing us closer to a life that is authentically ours.

In redefining our relationship with failure, we cultivate a mindset that sees beyond the immediate setbacks. We understand that each misstep is an opportunity to refine our goals, to realign them with our evolving values. This dynamic interplay between values and goals is what propels us forward, infusing our lives with meaning and purpose.

In this light, failing without shame becomes not only possible but empowering. It invites us to live courageously, to pursue our dreams with unwavering conviction, and to redefine success on our own terms. By embracing our personal values and setting goals that honor them, we chart a course towards a fulfilling and authentic life.

Creating a Life of Fulfillment

Imagine a life where every morning you wake up with a sense of purpose, where each day is filled with actions and decisions that align with your deepest values. This is not just a dream but a reality that can be achieved by shifting your perspective on failure. Often, we view failure as a wall that we slam into, but what if we saw it as a stepping stone instead?

The first step towards a fulfilling life is to redefine what failure means to you. Society often labels failure as a mark of shame, something to hide away and never discuss. However, by acknowledging failure as an integral part of growth, we open ourselves up to endless possibilities. Each setback becomes an opportunity to learn, adapt, and ultimately succeed in ways we never imagined.

Consider the notion that fulfillment is not about achieving perfection but about progress. Every attempt, every mistake, and every success is a brushstroke on the canvas of your life. By shifting your focus from the fear of failing to the joy of trying, you empower yourself to take risks, explore new horizons, and discover passions you never knew existed.

Creating a life of fulfillment also requires you to align your actions with your core values. Take a moment to reflect on what truly matters to you. Is it love, creativity, community, or adventure? When you let these values guide your decisions, you will find a profound sense of satisfaction in even the smallest achievements. This alignment transforms mundane tasks into meaningful experiences, making every day a step towards fulfillment.

It's also crucial to cultivate resilience. Life will inevitably throw challenges your way, but by building resilience, you develop the strength to weather any storm. This doesn't mean avoiding difficulties but facing them head-on with courage and determination. Resilience is the bridge between failure and success, allowing you to bounce back stronger each time you fall.

Moreover, fostering a supportive environment can significantly influence your journey towards fulfillment. Surround yourself with individuals who uplift you, who understand the importance of growth through failure, and who encourage you to pursue your dreams relentlessly. Their support will not only provide comfort in times of doubt but also inspire you to reach heights you once thought unreachable.

Finally, practice gratitude. In the pursuit of fulfillment, it's easy to focus on what you lack rather than what you have. By appreciating the present moment and acknowledging the abundance in your life, you cultivate a mindset of contentment. This gratitude fuels your journey, reminding you that fulfillment is not a distant destination but a state of being you can experience right now.

As you navigate this path, remember that creating a life of fulfillment is a personal journey. There is no one-size-fits-all blueprint, but rather a unique adventure tailored to your aspirations and dreams. Embrace failure not as an adversary but as a companion that guides you to a life rich with purpose, joy, and fulfillment. This is the essence of living without shame, transforming failures into the foundation upon which a truly fulfilling life is built.

Celebrating Small Wins

In a world that often glorifies grand achievements and monumental milestones, it's easy to overlook the significance of small victories. Yet, these seemingly minor triumphs are the building blocks of success and the fuel that keeps the engine of progress running. Recognizing and celebrating small wins is not merely a motivational tool; it's a vital practice for fostering resilience and maintaining momentum on the path to achievement.

When we acknowledge and celebrate small wins, we shift our focus from what we lack to what we've accomplished. This shift in perspective is crucial for maintaining a positive mindset, especially when facing challenges or setbacks. Every small win is a testament to our efforts, a reminder that progress is being made, and that we are moving in the right direction. It's the accumulation of these small victories that ultimately leads to significant accomplishments.

Consider the psychological impact of celebrating small wins. Each time we acknowledge a small success, our brain releases dopamine, a neurotransmitter associated with pleasure and reward. This chemical reaction not only boosts our mood but also reinforces the behavior that led to the win. As a result, we're more likely to continue engaging in productive actions, creating a positive feedback loop that propels us toward our larger goals.

Moreover, celebrating small wins helps to build confidence and self-efficacy. When we recognize our achievements, no matter how minor they may seem, we reinforce our belief in our abilities. This growing confidence empowers us to tackle more significant challenges with courage and determination. It's a powerful reminder that we are capable of overcoming obstacles and achieving our aspirations.

Incorporating the celebration of small wins into our routines also encourages a more mindful approach to life. It urges us to pause and reflect on our journey, appreciating the progress we've made rather than solely fixating on the destination. This practice of mindfulness enhances our overall well-being, reducing stress and increasing satisfaction with our efforts.

Furthermore, celebrating small wins fosters a culture of positivity and encouragement, both within ourselves and in our communities. When we share our victories, no matter how small, we inspire others to recognize and celebrate their own achievements. This collective recognition of progress creates an environment where success is not only expected but celebrated at every stage.

To truly embrace the power of small wins, we must create rituals and habits that encourage regular acknowledgment of our achievements. Whether it's through journaling, sharing with a friend, or taking a moment of quiet reflection, these practices ensure that we remain connected to our progress. By doing so, we cultivate a sense of gratitude and fulfillment that enriches our lives.

In the pursuit of our goals, it's essential to remember that success is not solely defined by the final outcome but by the journey we undertake to reach it. Each small win is a milestone that deserves recognition and celebration. By valuing these moments, we not only enhance our personal growth but also pave the way for more significant achievements. Let us cherish every step forward, recognizing that each small victory is an integral part of our grand tapestry of success.

The Ever-Evolving Path of Success

Success is not a static destination, but rather an ever-changing landscape that requires constant adaptation and resilience. In a world where the only certainty is change, the traditional definitions of success have become obsolete. We are no longer confined to the linear path of achievement that was once so heavily emphasized. Instead, we are invited to redefine success on our own terms, recognizing its fluidity and the personal growth it entails.

The notion that failure and success are polar opposites is a misconception that has long hindered progress. In reality, failure is an integral part of the success equation. It is through our setbacks that we learn, grow, and ultimately redefine what it means to succeed. This perspective shift allows us to approach challenges with curiosity rather than fear, transforming obstacles into opportunities for innovation and creativity.

In this new paradigm, success is not a solitary pursuit but a collaborative endeavor. It is about building relationships, nurturing connections, and fostering a community of support. The strength of our networks often determines our ability to navigate the complexities of our evolving paths.

By surrounding ourselves with diverse perspectives and constructive feedback, we enhance our capacity to adapt and thrive in an ever-shifting environment.

Moreover, the process of redefining success involves a deep understanding of our values and passions. It requires introspection and self-awareness, enabling us to align our actions with our core beliefs. This alignment is crucial as it fuels our motivation and sustains our perseverance in the face of adversity. When our pursuits are grounded in authenticity, we find a sense of fulfillment that transcends conventional measures of success.

As we navigate our unique paths, it is vital to remain open to change and ready to pivot when necessary. The ability to adapt is a powerful tool in our arsenal, equipping us to seize opportunities that arise unexpectedly. Flexibility and resilience become our allies, allowing us to re-calibrate our goals and strategies as circumstances evolve. This dynamic approach ensures that we remain relevant and capable of achieving success in its myriad forms.

Importantly, the redefinition of success requires a shift in our perception of achievement. It is not solely about reaching the pinnacle but about the journey itself—the lessons learned, the skills acquired, and the personal growth experienced along the way. Each step, whether forward or backward, contributes to our development and enriches our understanding of what it truly means to succeed.

In embracing the ever-evolving path of success, we liberate ourselves from the constraints of external expectations and societal pressures. We gain the freedom to chart our own course, driven by passion and purpose. This autonomy empowers us to pursue goals that resonate deeply with our personal aspirations, leading to a more meaningful and fulfilling existence.

Ultimately, success is a living, breathing entity that thrives on change and adaptation. By embracing its fluid nature, we unlock the potential within ourselves to achieve greatness, not by conventional standards, but by our own. This transformative journey, marked by resilience and self-discovery, is the true essence of success in an ever-evolving world.

ABOUT THE AUTHOR

My name is Tamika D. Roberson, and I proudly hail from St, Petersburg, Florida. As the only child of and extraordinary mother, I navigate life as a devoted mother to my loving son. My journey has not been without its challenges; growing up without a father left me grappling with deep seated pain and longing for love and affection. In my youth, I often felt insecure and shattered, yet I have emerged stronger, reclaiming my power as an adult.

In my journey of healing, I have learned the importance of nurturing the wounded girl within me. Embracing my pain and emotions has been crucial in transforming my experiences into strength. This journey ignites a profound desire within me to share hope with women who are suffering. I want to assure them that healing is possible.

My mission is to inspire and empower women to dream big and pursue their personal, professional and spiritual aspirations. Together, we can transform pain into purpose and create a brighter future. Through life-changing experiences, I discovered the undeniable power of encouragement. Therefore, I empower those around me to heal, love themselves and stop settling for less than they deserve. If you want to achieve anything in life, you need to be certain that you will achieve it. YOU MUST BELIEVE IN YOURSELF!

www.ingramcontent.com/pod-product-compliance
Lightning Source LLC
Chambersburg PA
CBHW071703210326
41597CB00017B/2309